Praise for *Future*

"This book is going to smash your existing views about anxiety—and replace them with more helpful ones. As an expert on the neuroscience of emotion, Tracy Dennis-Tiwary has delivered the riveting read we all need to help us learn to worry well instead of worrying less."

—Adam Grant, #1 *New York Times* bestselling author of *Think Again* and host of the podcast *WorkLife*

"If you're feeling more anxious than usual and, on top of that, feeling anxious about feeling anxious, then this book is for you. Anxiety, it turns out, is a feature, not a bug, of human nature. In *Future Tense*, Tracy Dennis-Tiwary offers a much-needed and compelling guide—based on years of scientific research and her personal clinical experience—to the emotion that is in some ways most easily misunderstood and, therefore, most undervalued. Clear, practical, and incredibly readable!"

—Angela Duckworth, PhD, Rosa Lee and Egbert Chang Professor, University of Pennsylvania, and *New York Times* bestselling author of *Grit*

"A fresh, hopeful approach to anxiety that will soothe readers facing a world filled with pandemics, war, and political turmoil."

—*Booklist*

"A powerful and deeply informed new voice in the important conversation around anxiety and its causes and effects. *Future Tense* offers knowledge, empathy, and clarity in these times when chronic emotional pain has been normalized. Framing how anxiety works in our favor is a revolutionizing shift in perspective."

—Alanis Morissette

"Reading *Future Tense* is an epiphany; it will turn your understanding of anxiety on its head and point you to new paths forward. It offers a long-overdue challenge to the medicalization and numbing of anxiety. Instead, Tracy Dennis-Tiwary encourages us to consider how to live and grow with anxiety and find creativity in dealing with life's fundamental uncertainties. A wide conversation around this book is urgently needed for our anxious times."

—Dacher Keltner, PhD, professor at UC Berkeley, faculty director, Greater Good Science Center

"*Future Tense* teaches us—with incredible research and great storytelling—that counter to everything we've thought and been taught, anxiety, when understood and used wisely, is one of the most valuable emotions to help us achieve our dreams. If you're prone to anxiety like me or live or work with people who are anxious, this is a must read!"

—Marc Brackett, director of the Yale Center for Emotional Intelligence and bestselling author of *Permission to Feel*

"*Future Tense* is groundbreaking. Filled with wisdom, compassion, and humor, it shatters our long-held assumptions and sets the stage for a new, hopeful way of understanding how to live—and thrive—with anxiety."

—Reshma Saujani, CEO of the Marshall Plan for Moms and founder of Girls Who Code

"Our minds classify anxiety as 'bad,' but that very idea keeps anxiety from delivering its often useful messages about what's ahead. It's time for the whole culture to learn how to use anxiety when it's helpful and let it go when it's not, but that starts with learning what it is and how to feel it. This wise and well-written book will help. Highly recommended."

—Steven C. Hayes, PhD, originator of acceptance and commitment therapy (ACT) and author of *A Liberated Mind*

"The problem isn't anxiety itself, but our beliefs about it and our attempts to avoid it, which are not only destined to fail but also to make us weaker and more fragile. It's a vicious cycle. Dennis-Tiwary suggests many ways of coping. . . . Anxiety brought [my daughter] to rock bottom, but by sitting with it, getting through not around, she's deeper, stronger, and so much wiser. Like Dennis-Tiwary says, it's horrible, it feels terrible—but it's beautiful, too."

—*The Guardian*

FUTURE TENSE

FUTURE TENSE

Why Anxiety Is Good for You
(Even Though It Feels Bad)

Tracy Dennis-Tiwary, PhD

HARPER

NEW YORK . LONDON . TORONTO . SYDNEY

HARPER

A hardcover edition of this book was published in 2022 by Harper Wave, an imprint of HarperCollins Publishers.

FIRST HARPER PAPERBACKS EDITION PUBLISHED 2024.

The Library of Congress has catalogued the hardcover edition as follows:
Names: Dennis-Tiwary, Tracy, author.
Title: Future tense : why anxiety is good for you (even though it feels bad) / Tracy Dennis-Tiwary, Ph.D.
Description: First edition. | New York, NY : Harper Wave, [2022] | Includes bibliographical references and index.
Identifiers: LCCN 2021062963 | ISBN 9780063062108 (hardcover) | ISBN 9780063062122 (ebook)
Subjects: LCSH: Anxiety.
Classification: LCC BF575.A6 D46 2022 | DDC 152.4/6--dc23/eng/20220118
LC record available at https://lccn.loc.gov/2021062963

ISBN 978-0-06-306211-5 (pbk.)

24 25 26 27 28 LBC 5 4 3 2 1

For Vivek, Kavi, and Nandini

Contents

from seeing this, and so we experience it only as something negative and disabling.

PART II HOW WE WERE MISLED ABOUT ANXIETY
Why and when our understanding of anxiety went wrong and how science and modern life exacerbate this misunderstanding

The psychological and medical sciences turned anxiety into an ailment. But even before that, medieval views of emotion and the life of the soul demonized anxiety. We learned to think of it as something to avoid and suppress, which only causes it to spiral out of control.

By thinking of anxiety as an unwanted affliction, we do everything in our power to do away with it. The widespread overprescription of anti-anxiety medications and painkillers is a prime example—one that has had damaging, even fatal, results.

Digital technology is a driver of unhealthy anxiety because it facilitates escapism and disrupts nourishing social connections. Blaming technology for all problematic anxiety, however, is a mistake—it misses the complexity of the problem while keeping us from seeing how to use digital technology in better ways.

PART III HOW TO RESCUE ANXIETY
Seeing anxiety as an ally will improve every part of our lives and can promote exceptional resourcefulness, creativity, and joy. In rescuing anxiety, we rescue ourselves.

Uncertainty puts us on edge. By leaning into that discomfort to figure out what to do about uncertainty—even in the middle of a global pandemic—we open doors to possibilities we never imagined before. When we do so, anxiety is the secret sauce.

When we accept the discomfort of anxiety and listen to what it's teaching us, we become more creative—whether we're creating works of art or figuring out what to make for dinner.

Too often, we react to our children's anxiety with accommodation and overprotection. We do so with the best of intentions and because we believe they are fragile, but we are wrong. We can help kids be their strongest and most resilient when we stop fearing their anxiety—and our own.

If you've read this far, you've changed your mindset about anxiety. It's time to do something about it.

FUTURE TENSE

Prologue

"Whoever has learned to be anxious in the right way," a famous philosopher once wrote, "has learned the ultimate."

Hang on—there's a right way and a wrong way to be anxious? That sounds to me like one more thing to be anxious about.

Yet Søren Kierkegaard, whom I like to think of as the Patron Saint of Anxiety, was onto something important.

You hate feeling anxious. I do, too. Everybody does. It's an emotion that can be distressing, burdensome, and debilitating. And because of that we're all missing the very thing Kierkegaard was getting at: anxiety wants to be our friend. It wants to be recognized and acknowledged and listened to and cherished and heeded. It feels terrible because it's trying to tell us something important that we'd rather not hear—as a good friend often does. Because if we do listen, our lives will be infinitely better than if we do what we really want to do when anxiety pays us a visit: run away and hide.

What's wrong with that? Isn't anxiety a personal failure, a sign that something is wrong with us and with our life, something to be fixed and eradicated? Yet nobody in the history of time ever eradicated anxiety—thank goodness, because that would be a disaster.

This book is the story of an emotion that is painful and powerful, terrible and funny, exhausting and energizing, and imperfect. It's like life. It's like being human. It *is* being human. If you read this book, I believe you will change the way you view anxiety. It's like the famous optical illusion Rubin's vase: look at it one way, and you see a vase, but relax your gaze and out pop two people in profile looking at each other across the vase-shaped space that separates them.

Making this paradigm shift, reclaiming anxiety as our friend and ally, isn't just going through a set of exercises and interventions. It's not just my telling you that anxiety sucks—although it really, really does at times—and describing twenty things you can do to feel better. Nor is it my telling you to glorify anxiety, or believe that you always need anxiety to be productive, creative, or to perform at your peak. You don't. Rather, it's creating a new mindset about anxiety—a fresh set of beliefs, insights, and expectations that allows you to explore anxiety, learn from it, and leverage it to your advantage. Achieving a new mindset will *not* fix anxiety itself—because the emotion of anxiety is not broken; it's how we cope with anxiety that's broken. Creating a new mindset is the best—and only way—we can repair that. This is the sole purpose of this book.

I hope Saint Søren would approve.

PART I

Why We Need Anxiety

What Anxiety Is (and Isn't)

Dr. Scott Parazynski and his space shuttle crewmates were speeding seventeen thousand five hundred miles an hour on their way out of Earth's atmosphere. Their destination was the International Space Station, a scientific hub, a stepping-stone for exploration of the solar system, and the largest structure humans have ever put into space. To many people, the ISS represents the pinnacle of human achievement.

By the time that mission took place in 2007, Scott was a veteran of four space shuttle flights and several extravehicular activities—space walks—in orbit. After retiring from NASA, he became the first person to have both flown in space and climbed to the summit of Mount Everest. This is a person who is comfortable with risk. But this mission carried an additional burden of significance. It had been delayed for three years after the space shuttle *Columbia* disaster, in which the spacecraft had disintegrated as it reentered the atmosphere, killing all seven crew members.

Yet for Scott and his team, the mission was worth the potential danger. They were to deliver and install a key component of the ISS that would connect and unify the US, European, and Japanese space labs within the station, providing additional power and life support and significantly expanding its size and capabilities.

After a week of new installations and routine repairs, things took an unexpected turn. Scott and a fellow crew member had just installed two huge power-generating solar cell panels. When the panels were opened and extended for the first time, a guide wire snagged, causing two large tears in them. That was a serious problem because the damage prevented the panels from expanding fully and generating enough energy to do their job.

For Scott to repair the torn solar cells, the team had to jury-rig an exceptionally long tether that would attach Scott to the end of a boom and then connect him—by his feet—to the end of the ISS's robotic arm. Dangling from the boom, it took him forty-five minutes to move ninety feet along the wing and reach the damaged panels. His skills as a surgeon were crucial as he painstakingly cut the snagged wire and installed stabilizers to reinforce the structure.

After seven nail-biting hours, the mission was a success. The crew on the ISS and the team back on Earth erupted into cheers as the repaired panels successfully expanded to their full length. A photograph of Scott seeming to fly above the glowing orange solar wing is an iconic image of intrepid exploration in space. His achievement is said to have inspired the death-defying spacecraft repair depicted in the movie *Gravity*.

Almost eight years after his celebrated feat, I had the immense pleasure of speaking with Scott on the stage of the Rubin Museum of Art's Brainwave program in New York

City. Tall, blond, and rugged, he looks like a circa-1950s American hero. He has the manners of one, too, with his easy smile and sincere humility.

I asked Scott how he had kept his cool that day with nothing but a space suit between him and the void. With the fate of the mission resting on his shoulders, what had been the secret of his success?

The answer? Anxiety.

Anxiety and Fear

I probably don't need to tell you what anxiety is.

It is a fundamentally human emotion, our companion since *Homo sapiens* walked upright. Anxiety activates our nervous systems, making us jittery and on edge, with butterflies in our stomach, a pounding heart, and racing thoughts. The word, derived from the Latin and ancient Greek words for "to choke," "painfully constricted," and "uneasy," suggests that it is both unpleasant and a combination of the physical and emotional—a lump in our throat, our body paralyzed with fear, our mind frozen with indecision. It wasn't until the seventeenth century that the word was commonly used in English to describe the range of thoughts and feelings we recognize today as anxiety: worry, dread, angst, and nervousness about situations with an uncertain outcome.

Often, you know why you are anxious: Your doctor calls, telling you she wants to schedule a biopsy. You are about to step out on stage to give a career-making speech before a crowd of five hundred strangers. You open a letter from the IRS informing you that it is auditing your tax return. Other times, our anxiety is more elusive, without any clear cause or focus. Like a maddeningly persistent alarm, this free-floating anxiety tells us that something is going wrong, but we can't find the source of the beeping.

Whether general or specific, anxiety is what we feel when something bad *could* happen but hasn't happened yet. It has two key ingredients: bodily sensations (unease, tension, agitation) and thoughts (apprehension, dread, worry that danger might be around the corner). Put the two together, and we see why choking gave anxiety its name. *Where should I go, what should I do? Will it be worse if I turn left or right? Maybe it's best if I just shut down or disappear altogether.*

Anxiety is experienced not only as a feeling in our bodies but also as a quality of our thoughts. When we're anxious, our attention narrows, we become more focused and detail oriented, and we tend to see the trees instead of the forest. Positive emotions do the opposite: they broaden our focus so that we get the gist of a situation rather than the details. Anxiety also tends to get our minds moving, worrying about and preparing for negative possibilities.

Though dread typically dominates our experience of anxiety, we are also anxious when we want something. I am anxious to board the plane that will take me to my much-overdue beach vacation, and no flight delays or rain better get in my way! This kind of anxiety is an excited *frisson* for a desired future. I am *not*, however, anxious to head to an annual holiday party, which is sure to feature the usual cast of characters drinking way too much. I already know I'll have a bad time there. But whether our anxiety is due to dread or excitement, we become anxious only when we anticipate and care about what the future holds.

So why isn't anxiety the same as fear? We often use the two words interchangeably, since both inspire unease and trigger "fight/flight" responses—the adrenaline rush, racing heart, and rapid breathing. Both anxiety and fear catapult our mind into similar states: laser focus, detail orientation, and readiness to react. Our brain is prepared, and our body is ready to snap into action. But there's a difference.

One day recently, I was rifling through an old box stored in the attic. My hand touched something warm and furry that moved. I jumped back faster than I would have thought possible and pushed the box away. Research on the human startle response shows that it took me only a couple hundred milliseconds to react. My heart was racing, I broke out in a sweat, and I was definitely more awake and alert than I had been moments before. It turned out that the creature in the box was a little field mouse.

My response to that mouse was fear.

Now, I'm not afraid of rodents. I think field mice are cute and an important part of the ecosystem. Yet my fear response didn't care that I don't expect mice to bite me. Fear wasn't interested in discussing the merits or cuteness of field mice and whether I really needed to jump back so quickly. And that's a good thing, because my automatic response would have come in handy had the critter in the box been a scorpion instead—just as reflexively pulling my hand away after touching a pot of boiling water protects me from getting burned further.

My fear was reflexive, much as it was for the little mouse as she darted around the box and then froze in the corner to avoid detection. At no point did I—or the mouse—feel anxiety about an uncertain future. Danger was in the certain present, and we both acted automatically and quickly to deal with it (although later I heeded my anxiety about letting a rodent run rampant in my house and relocated her to a neighboring field).

Of course, human emotional life is much more complicated than reflexive fear, anger, sadness, joy, and disgust. Emotion science identifies these as the basic, or primary, emotions. They're typically considered to be biological in origin and universal in expression. Animals share these emotions with us; that's how fundamental certain feelings are.

Then there are the complex emotions, including grief, regret, shame, hate—and anxiety. The basic emotions are the building blocks of the complex ones, which transcend instinct; they are less automatic and more amenable to our thinking our way out of them. I might feel anxious the next time I reach into a box in the attic, wondering whether I'll find another furry friend, but I can reassure myself that it is unlikely. Animals probably don't experience complex emotions such as anxiety in the way humans do; my little mouse doesn't have the capacity to vividly imagine a future in which giant hands might appear without warning to pluck her from the safety of her nest. If she did, that would make her the Jean-Paul Sartre of mice, complaining that

Hell is other mice as she retreated to her solitary box and grappling with existential angst as she waits for the next hand to descend. Whatever the case, what we can know for sure is that she will have learned through her encounter with me to *fear* hands if she ever sees them again, and her fear will end once she escapes to a warm, safe corner.

Fear is the immediate, certain response to a real danger in the present moment that ends when the threat is over. Anxiety is apprehension about the uncertain, imagined future and the vigilance that keeps us on high alert. It occurs in the spaces in between—between learning that something bad could happen and its arrival; between making plans and being helpless to take any real action—like fighting or fleeing, as animals do—to escape the danger. I can only wait to receive my biopsy results, to learn if the IRS examiner found any irregularities, or to hear whether my speech is followed by enthusiastic applause or half-hearted slow clapping. Anxiety exists because we know we are being slowly and inexorably pulled into a future that is either potentially unhappy or potentially happy. It's that uncertainty that makes anxiety hard to bear.

The Spectrum

Everyday anxiety is nothing surprising; we all experience worries, concerns, even moments of panic sometimes. But anxiety isn't a binary proposition, like a light switch that's either on or off. Imagine instead a dimmer sliding up and down, sometimes quickly, other times hardly at all. Low-level anxiety is present in our lives so often, like the air we breathe, that we might not even notice it. It happens when we open the door to meet our new boss or when we look outside and see snow coming down as we're packing up for our drive home; suddenly we're paying close attention to something we'd really rather not think about, but the feeling doesn't last for more than a minute or two. Once I meet my new boss, I soon get a sense of what she's like and my anxiety subsides. As I start driving home, I see that the roads are still clear, so my worries ease. Once we sense how things will turn out, our mild anxiety fades away like morning mist burnt off by the sun's warmth.

As we move along the scale from left to right, our anxious feelings get stronger, our focus turns into tunnel vision, and our worries really kick in. Let's take that prehistoric bugaboo, fear of the dark. It's not fear; it's anxiety. Unlike nocturnal animals, humans respond to darkness with apprehension about the unseeable hazards that *could* lie in wait. The search for light in the darkness is one of the most basic metaphors across human history. Even in prehistory, we can imagine, night lights—in the form of little fires?—were probably a hot commodity because we are so anxious about the dangers that hide in the darkness.

As we continue along the spectrum, one of the most common forms of moderate anxiety is the social kind—fearing the judgment and negative evaluation of others. What will the audience think of my speech? Will my employee evaluation go well? Will people laugh at my terrible dancing?

Even when we are confident in our abilities, many of us feel nervous before going out onto a stage. Sometimes, when we look out at our audience, all we can see is that one fellow falling asleep in the back. We don't even notice that everyone else is smiling and nodding in appreciation.

Over a matter of hours, even minutes, we might go from feeling mildly worried, then shifting up into high-intensity dread, before sliding back down the scale until we reach relief or even Zen-like calm. Even though high anxiety can feel out of our control, it's still just a point on a spectrum, so we can usually dial it back and return to our comfort zone.

That is because anxiety itself—the worry, dread, and nervousness; the distress over uncertainties; even the overwhelming panic—is not the problem. The problem is that the thoughts and behaviors we use to cope with anxiety can make it worse. When this happens more often than not, anxiety can start to lead us down the path toward an anxiety disorder. But the two—anxiety and anxiety disorders—are not the same.

The most crucial distinction between anxiety and an anxiety disorder is termed *functional impairment*—in short, when anxiety gets in the way of living life. The emotion of anxiety ebbs and flows, sometimes barely noticeable and sometimes distressing. But the disorder, by definition, involves more than temporary distress. For a person with an anxiety disorder, these feelings last for weeks, months, even years, and they tend to get worse over time. Most important, such feelings very often interfere with pursuing the things we treasure most, such as home life, work, and time with friends. This long-term impairment of our day-to-day activities and well-being is the sine qua non of anxiety disorders.

Take Nina. At thirty, she has built a career as a photographer, doing weddings and portraits. She has long known that she feels more comfortable watching people than being watched and being behind the camera instead of in front of it. Recently, however, her natural shyness has become hard to manage and has kept her from taking on new clients. She has started to believe that she appears bumbling, shaking, sweaty, and stupid to the world—and she wonders if that is indeed what she is. When she started failing to show up for work and struggling financially as a result, she decided to try therapy. As part of her treatment, the therapist asked her to take part in an experiment, which the therapist would record on camera.

First, Nina would pretend that the therapist was a potential client who was looking for a wedding photographer. She would talk with the woman as she would with any new client. During the conversation, she would also consciously do the things she typically does during interviews to handle her anxiety: look down and avoid eye contact while tightly clutching her camera or some other object to stop herself from shaking.

Then Nina and her therapist would reenact the interview with a key change: instead of looking down, Nina would consistently make eye contact and place her hands in her lap instead of clutching something.

Before starting the experiment, Nina's therapist asked

her, on a scale of 0 to 100, how much did she think she would shake? Nina thought 90. How sweaty would she appear, and how stupid would she sound? Again, Nina thought 90 on both for sure. She anticipated that she would be a nervous wreck, someone no one would ever want to hire to document a special day.

After acting out both versions of the conversation and viewing the recording, the therapist asked Nina: On a scale of 0 to 100, how did she actually look on camera—was she as shaky, sweaty, and stupid as she had anticipated? Nina was surprised to see that although she did seem nervous during the first part of the experiment, she didn't shake at all or seem to be sweating, and she sounded fine—maybe not brilliant but certainly not stupid. When Nina watched the second half of the experiment, when she made eye contact and wasn't clutching her camera, she couldn't help but notice that she suddenly appeared every inch the confident professional. She smiled, was well spoken, and offered good ideas and suggestions.

It's not that Nina wasn't feeling nervous. She was. But once she stopped acting like a wreck—looking away and holding on to the camera for dear life—she felt much less like one. That was because she stopped relying on ways of coping that unintentionally made her anxiety worse.

If changing a few key behaviors and perceptions can indeed help to alleviate painful, even debilitating anxiety, why are anxiety disorders the single most common mental health problem today? Why are they arguably on the rise, fast becoming the public health crisis of our era?

If that sounds like an overstatement, consider the statistics. A large epidemiological study conducted at Harvard, using a combination of diagnostic interviews and assessment of life impairment, showed that almost 20 percent of adults in the United States—more than 60 million people—suffer from an anxiety disorder every year. About 17 million people each year suffer from depression, the second most common mental health problem, and nearly half of them are also diagnosed with an anxiety disorder. Over a lifetime, the number of Americans who will suffer from one or more anxiety disorders jumps to a shocking 31 percent—more than 100 million of us, including teenagers and kids. Many seek therapy, but fewer than half show lasting change, even when receiving gold-standard treatments such as cognitive behavioral therapy. Women are disproportionately affected; almost twice as many women as men will be diagnosed with an anxiety disorder in their lifetimes.

Nine different anxiety disorders are diagnosed in the United States, not including trauma-related disorders such as post-traumatic stress disorder (PTSD) and compulsive disorders such as obsessive-compulsive disorder (OCD). Some anxiety disorders, such as phobias, primarily involve avoiding feared objects and situations, such as hemophobia, fear of blood, and claustrophobia, fear of being in a closed space. Other types of anxiety disorders involve intense bodily signs of fear, such as a panic attack, when a sudden outbreak of shaking, sweating, shortness of breath, chest pains, and a feeling of impending doom mimic what many of us think a heart attack probably feels like. In other types, such as generalized anxiety disorder (GAD), worries consume time and attention, causing people to avoid situations they used to enjoy and making it difficult for them to focus and perform at work.

Imagine the experience of Kabir, who first showed signs of intense anxiety when he was fifteen. At first, he feared only speaking in class. For days before a presentation he worried constantly, didn't sleep, and refused to eat. He made himself sick with worry. As a result, as time went on, he missed more and more days of school and his grades suffered. Soon, this extreme and constant worry emerged even about nonschool situations, such as when he was invited to a party or when he was to participate in a swim meet. Within months, he stopped doing both and broke off the few friendships he had. By the end of the year, he was having full-blown panic attacks, with heart palpitations and feelings of suffocation so extreme that he was convinced that he was having a heart attack.

By diagnostic standards, Kabir went from feeling highly anxious to developing social anxiety, GAD, *and* panic disorder. Whatever the labels, he was diagnosed *not* because he felt intensely anxious and worried but because he could no longer go to school, participate in activities, or keep friends. His way of coping with worry and anxiety had gotten in the way of his ability to live his life.

The key problem for people diagnosed with an anxiety disorder is not just that they experience intense anxiety; it's that the tools they have at their disposal to turn down the dial on those feelings are not effective—as was the case with Kabir, who coped with anxiety by eating and sleeping poorly, staying home from school, dropping out of sports, and isolating himself from his friends. Such attempted solutions serve only to avoid or suppress anxiety and just end up making it worse. In other words, although anxiety is fundamentally a useful emotion, the symptoms of anxiety disorders are worse than useless; they actively get in the way.

So it's not that we're in the midst of a public health anxiety crisis; we're in a crisis of the way we cope with anxiety.

Think of anxiety like a smoke alarm, warning that our house is on fire. What if instead of running out of the house and calling the fire department we just ignored the alarm, removed the battery, or avoided the places in the house where the alarm was loudest. Instead of listening to the critical information the alarm is giving—where there's smoke, there might be fire!—we imagine that it's not there. So instead of benefiting from the alarm and putting out the fire, we just hope and pray that the house doesn't burn down. That's not to say it's always easy to listen to anxiety. Intense, enduring anxiety can overwhelm our ability to perceive the useful information it might hold for us. Or, conversely, we fail to listen to it because we've decided that the only way to get things done in life is to suffer through regular, anxiety-fueled adrenaline rushes. Yet, when we believe that our anxiety is worth listening to, when we investigate it rather than revile it, we break such unhealthy cycles and come to realize that some ways of responding to anxiety turn down the dial on the continuum, whereas other ways—especially ignoring it—rev us up to unmanageable levels. Before we know it, our house *is* ablaze.

Of course, it's not just difficulties with coping that lead to debilitating anxiety. In many cases, experiences of chronic and unrelenting stress and adversity play a huge role. Sometimes life just doesn't let up, and any of us in such a situation would feel intense and overwhelming anxiety. Yet to say we're in the midst of a crisis of how we cope with anxiety does not negate that fact, because no matter what the cause, being able to cope with anxiety differently is part of the solution. And listening to our anxiety—believing that there can be wisdom in what it tells us—is the first step toward finding that solution.

Believing that our anxiety is worth listening to might be easier than we think. Imagine that you are running for president of a political organization. Your task is to give a campaign speech. You have three minutes to prepare your remarks, after which you will deliver a three-minute speech. You will be speaking in front of a panel of judges, and your performance will be videotaped and compared to videos of other candidates' talks.

If you are diagnosed with social anxiety, you live in fear of how other people will judge you. You're already very tough on yourself; even trying to think of your positive qualities makes you uncomfortable. So this entire experience sounds like torture.

As the judges watch you, they do nothing but frown, cross

their arms, shake their heads, and display other discouraging nonverbal feedback. After what feels like an eternity, your speech finally ends. Certainly, you've earned a break. But your trials are not over yet.

Now you are told to perform a tricky math problem in front of the same panel of judges: you must count backward from 1,999 by 13, out loud, as fast as you can. The evaluators call you out every time you pause, saying "You're counting too slowly. Please speed up. You still have some time left. Please continue." Every time you fail, someone says, "Incorrect. Please begin again from 1,999." Even those of us who are confident in our math skills would be rattled.

This mini–torture session is actually a famous research task called the Trier Social Stress Test, or TSST. The experiment was developed more than forty years ago; it creates stress and anxiety in almost everyone but is an especially painful experience if you struggle with social anxiety—your heart will pound, you will breathe faster, you will feel butterflies in your stomach, and you will stumble over your words. It would be reasonable to assume that these signs show that you are not coping with the challenge very well.

But what if before doing the TSST, you were taught to anticipate your anxious responses and were informed that they are, in fact, signs that you are energized and preparing to face the challenge ahead. You're informed that anxiety evolved to help our ancestors survive by delivering blood and oxygen to our muscles, organs, and brains, so that they work at peak capacity. And just in case you're not convinced, you read some impressive scientific studies that document proof of the numerous positive aspects of anxiety.

If you had learned all this before you underwent the dreaded TSST, would it have made a difference in how you handled it?

In 2013, researchers at Harvard answered that question. Their work showed that if socially anxious participants got the lesson about the benefits of anxiety, they reported feeling less anxious and more confident. The difference in their physiological response to anxiety was even more striking. Typically, when we experience high anxiety and stress levels, our heart rates increase and our blood vessels constrict. Once the research participants perceived their anxious bodily reactions as beneficial, however, their blood vessels were more relaxed and their heart rates were steadier. Their hearts were still pounding—the TSST is a strain no matter what you do in advance—but their cardiac patterns were more similar to healthy patterns when we are focused and engaged—when we are bravely meeting challenges.

This study showed that just by changing what we believe about anxiety—that it is a benefit rather than a burden—our bodies follow suit and believe it, too.

The Problem and the Solution

In this era of pandemic, political polarization, and catastrophic climate change, many of us feel overwhelmed by anxiety for our future. To cope, we have learned to think of the emotion as we do any other ailment: we want to prevent it, avoid it, and stamp it out at all costs.

As scientists become more aware of anxiety than ever before, why aren't the prevention and treatment of debilitating anxiety—anxiety disorders—keeping pace with those of physical diseases? Clearly, hundreds of books, thousands of rigorous scientific studies, and thirty different antianxiety medications are not helping enough. Why have we mental health professionals failed so spectacularly?

The fact is, we have it backward. The problem isn't anxiety; the problem is that our beliefs about anxiety stop us from believing we can manage it and even use it to our advantage—just as the participants in the TSST experiment learned. And when our beliefs make our anxiety worse, we are at greater risk of traveling down the path toward debilitating anxiety and even anxiety disorders.

When Scott Parazynski walked out into the void of space, laser focused and determined, it was anxiety that readied him for the worst. It enabled him, even before the mission began, to prepare for a perilous moment that he didn't know for sure was even coming. But he knew that an unhappy outcome was possible, as was a triumphant one, and so he trained for months, sharpened his skills, and cemented the trust he shared with his team.

Anxiety can be hard, disruptive, sometimes terrifying. At the same time, it can be an ally, a benefit, and a source of ingenuity. But to shift our perspective, we'll have to break down and rebuild our story of this emotion. This will require a journey, from the halls of academia to the theaters of the world, from medieval sermons of hellfire and brimstone to life during lockdown, from the infinite scroll of our cell phones to our kitchen tables.

If anxiety is such a great thing, why does it feel so bad?

2

Why Anxiety Exists

I'm in my car, paused at a traffic light. It turns green and I begin moving, when suddenly the driver parked to my left pulls out a few inches ahead of me but far enough to block my way. I lean on the horn, but he keeps moving, and so do I, until I reach the point where I care more about my paint job than whether the guy is going to cut me off. As he pulls ahead of me, I scream some choice words and send him dagger looks.

I didn't just feel vaguely annoyed. I was furious; morally outraged, even. My heart was pounding, I could feel the blood racing through my veins, and my face was set in a scowl. My body was humming with energy and ready to spring into action—even if that action consisted only of angry yelling.

I didn't like how those changes felt. They made me stressed, and I was ashamed that I couldn't just rise above it all. But my anger banished those intellectual musings by doing exactly what it was designed by evolution to do: make me fierce.

It's worth noting that the other driver wasn't really making much of a difference in my life; it was only a matter of being one car length ahead or behind. But it mattered a lot to my instinctive emotions. Anger prepared me to act forcefully—*just in case*. Luckily, we humans can usually mitigate our reactions to fit the situation—an ability that makes or breaks civilized society.

Almost by definition, anxiety and other negative feelings, such as anger, feel bad. It's a good thing they do.

More than 150 years ago, Charles Darwin came to the same conclusion.

The Logic of Emotion

Throughout human history, negative emotions have gotten a bad rap—irrational at best, destructive at worst. The ancient Roman poet Horace wrote, "Anger is a short madness." But in just the past 150 years, we have come to learn that emotions such as fear, anger, and anxiety are *not* merely dangerous; they can also be advantageous. Emotions are tools for survival, forged and refined over hundreds of thousands of years of evolution to protect and ensure that humans—and other animals—thrive. Indeed, from the perspective of evolutionary theory, emotions embody the logic of survival.

Darwin's earliest research was in geology and the extinction of giant mammals, which he conducted as an adventurous young man surveying coastal South America aboard HMS *Beagle*. That work of observation across regions of the uncharted Southern Hemisphere made him a star of the scientific community and gave birth to his first ideas on evolution. But it would be forty years later, in *The Expression of the Emotions in Man and Animals*, the third and final book of his trilogy on evolutionary theory, that he would apply his insights to the great terra incognita of the human mind: the emotions.

He had already explained principles of evolution in *On the Origin of Species*, and argued that humans and primates share a common ancestor in *The Descent of Man, and Selection in Relation to Sex*. Now, in *The Expression of the Emotions in Man and Animals*, he viewed emotions as he did any other universal traits found in animals: webbed footing, the shape of a tail, the color of fur or feathers. They had evolved as advantageous adaptations to environmental pressures over long periods of time. They were retained and passed on to future generations if they benefited the species. In other words, they contributed to the survival of the fittest.

Emotions meet the criteria of being advantageous adaptations. Take, for example, two animals squaring off in a fight over food. As they prepare to lock horns, literally or figuratively, their intense feelings prompt a repertoire of bodily reactions. When an animal's back arches and its hair stands on end, it appears larger and stronger. When it bares its teeth, furrows its brows, makes fierce noises, or brandishes its horns, it signals to the other animal that fighting such a strong adversary may not be worth it. These signals—displays of aggression—directly improve the chances that the other animal will withdraw, thus preventing violence and averting potential injury or death. Sending these archetypal signals benefits the species, as does the ability to interpret these messages. It's a win-win.

Darwin argued that if the actions connected to emotions are useful, they will be repeated and eventually become habits that can be inherited by offspring. He called this the *principle of serviceable associated habits* and noted, "It is notorious how powerful is the force of habit. The most complex and difficult movements can in time be performed without the least effort or consciousness." It is through the force of habit that facial expressions associated with emotions first evolved. For example, the furrowed brow of anger prevents too much light from entering the eyes, an important adaptation if one is in the midst of a struggle and can't afford to have one's vision obscured. In contrast, raising brows and opening the eyes wide increase the field of vision, which is desirable when fearfully scanning one's surroundings. The wrinkled nose and puckered mouth of disgust limit the intake of potentially rotten or poisonous substances. These reactions were useful—functional—and were thus enacted every time certain emotions arose.

In other words, actions learned through trial and error that lead to pleasure or avoid pain are adopted for future use because they are beneficial and help the individual survive. This idea is the bedrock of modern-day behavioral science, heavily influenced by Darwin, called the *law of effect*: the more an action leads to a good outcome, the more we do it.

These feeling-equals-action repertoires, such as "fear—open eyes wide" and "fight—show strength," are adaptive and useful, but they also achieve something else: they have direct effects on our nervous system. Darwin wrote, for example, "A man or animal driven through terror to desperation, is endowed with wonderful strength, and is notoriously dangerous in the highest degree."

These effects happen extremely quickly and automatically, which makes them valuable for promoting survival. They don't require time or forethought or even much energy. They just happen. It's a good thing, too, because in a split second we can protect ourselves, say by reflexively jumping out of the way of danger while simultaneously opening our eyes wide to take in as much information as possible about what's going on around us to decide what our next steps should be.

Another immense advantage of emotions is that they are social signals, conveying crucial information to other individuals of one's species or fellow members of one's tribe. Indeed, humans and other animals are biologically primed to pay attention to the emotions of our social partners as they react to us, such as the difference between someone looking at us lovingly and approvingly versus angrily or with disappointment. Even human infants freeze, alarmed, when they observe fear on an adult's face. They sense danger.

In a classic psychology experiment called "the visual cliff," a baby sits on one end of a bridge made of clear Plexiglas placed four feet above the ground. From the baby's angle, the Plexiglas is invisible—the baby sees just a long drop to the floor. On the other side of the invisible bridge sits the baby's mother. If she smiles and gestures for the baby to join her, almost all babies crawl right over the edge—that is, right into what looks to them like empty space. But when the mother expresses distress or fear, the baby stays put.

Why Anxiety Has to Feel Bad

Darwin caused a seismic shift in how we understand the role of emotion in our lives. Now, rather than being depicted as irrational and harmful, emotions—even negative ones—are seen as adaptive and useful. The trick is to master our emotions and be able to wield them as tools.

Functional Emotion Theory takes this premise as a starting point. It boils emotions down into two dynamic parts: appraisal and action readiness. This notion is very similar to Darwin's feeling-equals-action repertoires, positing that emotions inform and motivate us to do all sorts of useful things, such as overcome obstacles, build strong communities, and seek safety.

The first component, *appraisal,* is our perception of whether a situation is desirable—that is, does it allow us to get what we want or avoid what we don't want, both of which feel good? This sounds selfish and hedonistic, but from an evolutionary perspective, pursuing what feels good tends to promote our well-being and survival. My almost road rage, for example, involved the appraisal that the other driver was blocking my ability to get what I wanted: to move forward and get home. Moreover, because I perceived his actions as rude and unjust, he even obstructed my desire to live in a civilized world of considerate people.

We should keep in mind that the evolution of emotion was likely completed long before we humans developed dangerous addictive substances and other feels-good-but-is-clearly-bad-for-you things. In those cases, hedonism isn't such a useful benchmark.

Because appraisals are interpretations of situations vis-à-vis their relevance to our well-being, they provide information that directly informs the second component of emotion, *action readiness*—our reflexive responses that make us act in ways that get us what we want. So when my desires were thwarted by the other driver, my face, body, and mind revved up. Blood pumped faster through my veins, my attention was laser focused, and I sent facial signals of *Don't mess with me*. If he had hesitated from pulling out in front of me, I would have sped right past him. If he had gotten out of his car to scream back at me, God forbid, I had no doubt in the moment that I, all five feet, four inches of me, would have gotten out of my car, believing I could take him on. Putting aside whether that was an accurate assessment or a wise thing to do—it wouldn't have been—my anger gave me a fighting chance. Pun intended.

From a functional perspective, anxiety is a fascinating emotion because it acts a lot like fear but contains qualities of hope. Like hope, anxiety involves appraisals about an uncertain future. As a result, it's a protective alarm bell, triggering discomfort and apprehension about the possibility of future threats. But it's also a productive signal, telling us that there is a discrepancy between where we are now and where we hope to be and that averting threats and achieving our goals will require effort. As a result, anxiety activates action readiness tendencies to take flight or fight while simultaneously pushing us to work hard and achieve to get what we want but do not yet have. Like hope, anxiety cultivates endurance.

When our backs are against the wall, few other emotions keep us trained on the future so effectively, energizing and driving us to reach our goals, despite exhaustion or overwhelming obstacles.

Anxiety works so well *not* because it feels great to be anxious; just the opposite: it succeeds because it makes us feel so bad. Nervous. Worried. Tense. We'll do practically anything to make the feeling go away. This is called *negative reinforcement*—stopping the anxious feeling *is* the reward. Anxiety drives us to do things that protect us and motivate us toward productive goals, which then in turn, by reducing our anxiety, signals to us that our actions have succeeded. This makes anxiety, with its own built-in self-destruct system, one of our best survival mechanisms.

If we think of anxiety—and other unpleasant emotions—only as something to be squashed and controlled, we miss the fact that anxiety is fundamentally information. Imagine you've been sitting with free-floating anxiety for a couple days. You've been trying to ignore it, just keep calm and carry on, but it's getting to you. So you decide to tune in to what your anxiety is telling you. You go through a mental checklist: What's been bothering me? Is it the fight I had with my husband? No, that got resolved. Is it the work deadline looming over me? No, that's well enough in hand. Is it that my acid reflux has gotten a lot worse and I've been having stomach pains for the past five days straight? Ah, *there* it is. Bingo.

Once you identify the source of your anxiety, you have useful information. And you now know what actions to take. After you schedule an appointment with your doctor, your anxiety immediately begins to lessen. You're on the right track. When you later see your doctor and get a good plan to solve the problem, your anxiety disappears. Mission accomplished. Anxiety has done its job.

However, if you were to find out that there was actually something seriously wrong with your health, your anxiety would return—and motivate you to take whatever additional steps were necessary to deal with the illness. Without anxiety, you might have lost the chance to survive and thrive.

So anxiety *must* feel bad, must always have at least a tinge of the unpleasant—so that it demands our attention, informs us, and motivates us to act if for no other reason than to get relief from the anxiety itself.

That's not to say that anxiety always leads us to good and helpful actions. It might drive us toward unhealthy obsessiveness. Or the opposite, we might choose to ignore it—to procrastinate, self-medicate, or do other unhelpful things intended to merely silence the emotion. Yet, had we human beings succeeded in smothering anxiety over the course of our evolution, the loss of this important emotion could have been catastrophic.

Try to imagine prehistoric humans without anxiety, thinking mostly of the present, never bothering to worry or dream about the future as long as their bellies are full and their bodies are comfortable. Without anxiety, we might have died out as a species long ago. We certainly would never have become animals who are capable of scientific and technological advances, who have traveled into space, and who have created artistic works of transcendent beauty. Why would we bother? We would just take life a day at a time, motivated to feel neither apprehension, excitement, wonder, nor hope. In this sense, anxiety emerged from the fires of evolution to drive us to the pinnacle of our humanity. Only those who care to look past the present and think about the future can build civilizations.

The Emotional Brain

Evolutionary theory helps explain why some emotions are shared by most animals and others seem uniquely human. We can detect what seems to be fear in mice, identify signs of loss and sadness in elephants, dogs, and primates, and interpret anger in the ferocity of predators. As Darwin wrote, quoting William Shakespeare's *Henry V*:

> *But when the blast of war blows in our ears,*
> *Then imitate the action of the tiger:*
> *Stiffen the sinews, summon up the blood,*

Then lend the eye a terrible aspect;
Now set the teeth, and stretch the nostril wide.

Emotions such as fear presumably evolved among our pre-mammalian ancestors. Their brains also had the structures responsible for detecting and responding to threats that are involved in human fear. The same is true of aggressive and defensive emotional responses linked to areas such as the hypothalamus, which controls key bodily functions by activating the fight/flight, or sympathetic, nervous system.

On the other hand, affiliative emotions such as love of offspring are more likely to have evolved to support the survival of mammals, which require extended care throughout their helpless infancy (unlike other animals, such as reptiles and amphibians, which leave their offspring even before birth, or birds, which often kick their babies out of the nest once they can fly). More elaborate social emotions, such as guilt and pride, tenderness and shame, appear to have evolved only among social primates—humans and perhaps the chimpanzees and great apes—since emotions such as these, which keep us beholden to our tribe, come in handy to deter bad or sociopathic behavior.

We think of fear as being an ancient, more primitive form of anxiety, rooted in brain structures such as the amygdala, which is part of our limbic, or "emotional," brain. But the amygdala—actually, we have two of them, and the word comes from the Greek word for "almond," *amygdalē*, because of their shape—is much more than a fear center; it's a central hub connecting the sensory, motor, and decision-making areas of our brains. The amygdala is activated when we are fearful or anxious, but it also alerts us to salience, novelty, and uncertainty—anything unusual that might affect us. When faced with something new or ambiguous, such as someone looking at us with a hard-to-read expression, our amygdala is activated. But when we receive a reward, our amygdala also fires up. That's why it's not just for negativity; it is the brain center that helps us navigate the push and pull of fear and desire. It's considered a central part of the reward system of the brain, powerfully shaping what we learn and remember about the good and the bad, as well as what we decide to do about them.

A key neurotransmitter underlying the reward system—and anxiety—is dopamine. Dopamine's job is to ferry information to and from the reward system to other areas of the brain involved in things such as decision making, memory, movement, and attention. It's often described as the "feel-good hormone" because it's released when a person does something that brings pleasure, such as ingesting food, taking drugs, having sex, or viewing Instagram. But spikes in dopamine don't just follow something rewarding, they also precede it, activating brain areas that motivate us to pursue those rewards. That's why even though dopamine doesn't actually make us feel pleasure as other hormones such as endorphins do, it is strongly implicated in addiction.

Researchers are now discovering that it's not just addictive, pleasurable things that trigger the release of dopamine. Anxiety does, too. Why? Anxiety motivates people to pursue good, rewarding outcomes and avoid bad, punishing ones. Dopamine is released when a desirable outcome is obtained, as well as when we feel the relief of anxiety consequently decreasing. The relief of dopamine signals both of these pleasures, teaching us that doing something with anxiety is good and motivating us to keep taking effective action when anxiety arises.

Anxiety successfully integrates the limbic brain–based fear and reward systems, but it wouldn't truly be anxiety without the contribution of the more recently evolved outer layer of the brain, the cerebral cortex. One part of the cortex, the prefrontal cortex (PFC), is active when we invoke *executive functions*, such as inhibition of actions, attention control, working memory, and decision making. These functions are constantly recruited and activated during experiences of anxiety in order to direct and regulate every facet of our emotional response—the appraisals, the action readiness tendencies, and the feeling of emotion. The amygdala also communicates with areas of the brain that allow us to draw on memories and thoughts to make sense of our anxiety and put it into the context of who we are and what we care about—areas such as the hippocampus, which supports learning and long-term memory, and the insula, which is involved in consciousness and self-awareness.

In other words, although the amygdala is a central component of the emotional brain, it does not exist in a vacuum; it's an interconnected hub within a web of brain areas and the abilities they support. This is what we mean by *neural network*. The more recently evolved areas of the brain, such as the PFC, regulate our older, limbic brain, including the amygdala. The PFC is slower and more deliberative, while the amygdala is faster and more automatic as we negotiate a world full of danger, reward, and uncertainty—things we need to pay attention to in order to survive.

So it goes with anxiety; it doesn't arise just from the automatic, reflexive, ancient fear brain. Neither can it be traced simply to the evolved, deliberative, cognitively sophisticated cortex. It's the intersection and the balance between the two.

Anxiety and the Biology of Threat

Key to the neuroscience of anxiety is the defensive brain, a coordinated network of regions that work together to detect real or potential threats and coordinate our efforts to defend against danger. This includes the areas of the brain discussed above, such as the amygdala and PFC, as well as structures such as the periaqueductal gray (PAG), which helps control automatic fight/flight behaviors.

This defensive brain network allows us to learn and remember about threats quickly and effortlessly. If you're bitten by a dog on Monday, your defensive brain responses will be more quickly activated when you see that dog—or any dog—on Thursday. These responses make us nervous and prepare us for the potential of another bite. They are also a foundation for learning: we learn to be more cautious around dogs and to detect signs that they might be aggressive. The benefit is obvious.

But this defensive advantage can become too much of a good thing. When fear of dogs becomes an anxiety disorder—cynophobia—we start to overestimate the danger any dog poses. If we can't tell the difference between a snarling junkyard defender and a sweet little puppy, our signals for threat and safety are mixed up, like crossed wires. We exaggerate potential hazards, feel constantly on guard, and end up exhaustively scanning our environment, trying to make sense of why our internal alarm bells are still going off.

When this happens, something psychologists call the *threat bias* can develop. This is an unconscious habit of seeing the world through the lens of negativity—being constantly on the lookout for threat or danger, getting stuck on negative information when we do detect it, and ignoring evidence that we're actually safe and sound. The threat bias, in other words, is like an information filter that favors negativity over safety.

Imagine this: You're giving a speech in front of a hundred people. You look out into the crowd and immediately fixate on the one audience member who is frowning or, God forbid, falling asleep. In an instant, you develop tunnel vision for that person as if no one else exists. You don't notice that the ninety-nine other people are listening attentively, smiling, and nodding. This attentional spotlight on the negative audience member is the threat bias. The result is that you stay on high alert for other negative feedback, ignoring all evidence that you're doing a good job. In the moment, however, you're not aware of this. You just know that you're nervous and about to fail.

Like other biases, the threat bias is an evolved heuristic, a quick and automatic yardstick for the brain to measure what's going on in our lives. It piggybacks onto our instinctive ability to quickly and automatically detect threats, which is the core mission of the defensive brain. But the threat bias causes an imbalance in what we pay attention to, so that we prefer to see negativity at the expense of the positive. When the threat bias becomes a habit, it puts our fight/flight response on a hair trigger and skyrockets our feelings of anxiety.

The example of faces in a crowd is a telling one, because our brain's response to human faces is a key aspect of the threat bias. Faces are among the most compelling things our brains encounter. Within milliseconds, we reflexively identify and decode the subtlest facets of a facial expression. We couldn't stop ourselves if we tried. There's even a place in our brains that is specialized for this job: the fusiform face area. Darwin predicted this long ago; humans who survived and thrived (and therefore passed along their genes) were those who could decode human faces. Some faces are particularly strong attractors to our brains; we pay extra attention to angry expressions, for example, because they signal danger. But when we're chronically anxious, our ability to judge what's dangerous and what's not can become distorted.

Research from my lab and others has revealed that understanding the threat bias can help us predict whether healthy anxiety will veer down the path toward anxiety disorders. The most important thing is not whether our attention is grabbed by negativity; it's what we do with that information. Do we stare down at our speech notes, never looking up? Or do we look out into the audience to find the smiling faces? In other words, do we use anxiety to direct our attention to reward?

Imagine sitting in front of a computer screen, looking at a series of faces, some angry, some happy, and some neutral. It's a deceptively simple task, but put some highly anxious people on the spot, and you learn a lot. Using eye tracking, which follows our gaze, and electroencephalography (EEG), which measures how our brains respond to faces, we have documented the threat bias—that a large proportion of anxious people pay *too much* attention to the threatening angry faces. And the most severely anxious among them also pay *too little* attention to the happy faces—just like the person giving the speech in front of the audience. How and whether we draw on one of the richest sources of positivity and reward—our supportive social connections—has an immense impact on our anxiety.

The Social Brain and Anxiety

Being with loved ones relieves anxiety. This makes intuitive sense, but what is happening under the surface to make it so?

Anxiety orients us toward others by changing the chemistry of our bodies. It ramps up our levels of the stress hormone cortisol. Anxiety also triggers our brains to produce oxytocin, known as the *social bonding hormone*. This chemical is all about connecting with others; it's the hormone that's released when we are in love, and when women have babies, it helps not only with the process of birth but with emotional bonding with the newborn. Oxytocin makes us yearn for those we care about. So by stimulating its release, anxiety encourages us to connect with others.

Add to that the fact that oxytocin has direct antianxiety effects on the brain. Studies have shown that increasing levels of oxytocin in the blood cause stress hormone levels to plummet and amygdala activity to decrease—just like when you take antianxiety medications such as benzodiazepines. So powerful are the effects of oxytocin that researchers have begun to examine potential therapeutic uses of it in the treatment of anxiety disorders.

Now that we're connected and our brains are biologically soothed, how does being with loved ones relieve anxiety in more observable ways? Back in the early 2000s, a simple but astute clinical observation inspired some new ideas about this. A psychologist was conducting therapy with a military veteran suffering from post-traumatic stress disorder. For years the veteran had refused to seek treatment, saying he didn't need to see a shrink. But his wife, who accompanied him that day, finally convinced him to try. The patient slowly and haltingly shared his painful memories of combat. Each time he got upset and wanted to leave the therapist's office, his wife reached over and gently took his hand. Every time she did so, he was able to keep talking, working through his trauma. He eventually bene-fited from therapy.

That experience got the therapist, who was also a clinical neuroscientist, thinking in a different way about the impact of social connection on anxiety. A few years later, in 2006, he and his colleagues at the University of Wisconsin put the idea to the test. They recruited volunteers to take part in a study and then gave them something very concrete to be anxious about: unpredictable electric shocks administered in a magnetic resonance imaging (MRI) machine.

The potential to be shocked is threatening enough, but being in an MRI machine, a large tube surrounded by a giant superconducting magnet, made it even scarier. The partici-pants lay on a table and were transported into the machine, which, all the while, was making a tremendous racket, like a constant, rapid hammering.

One-third of the participants entered the scary, loud, claustrophobic machine alone. The rest were allowed to hold either the hand of a stranger or that of their romantic part-ner. Those who held the hand of their romantic partner showed the lowest levels of activity in brain regions associ-ated with anxiety: the amygdala and a specific area of the pre-frontal cortex related to the management of emotions, the dorsal-lateral PFC. But that was true mainly for those who reported a high level of relationship quality. Those with a lower-quality relationship showed much more anxiety-related brain activity *and* higher levels of stress hormones than did the satisfied hand-holders. Those who held hands with a stranger had higher anxiety-related brain activity, involving even more regions, such as the anterior cingulate cortex. And the final group, those who faced the threat of shock alone, with no hand to hold? They showed the highest levels of brain activation in all regions. Their brains were working very hard to manage their anxiety.

This study illustrates how social connection, even if it's superficial, can relieve anxiety. The mere presence of other people, particularly loved ones, helps our brains handle the stress of threat. It's called *social buffering*: because humans evolved in groups, we learned early on to rely on one another so that we expend less emotional energy and reap more benefits by facing difficulties together rather than alone. Every challenge becomes harder when we are socially isolated.

An extreme case in point is solitary confinement in prison. In the US, the practice was introduced by Quakers in the early nineteenth century to provide time and space for prisoners' self-exploration and penance. Soon, however, they observed the alarming disintegration that we see today: prisoners banging their heads against walls, cutting themselves, attempting suicide. The Quakers soon ended the practice (though we have not). So basic is our need for social connection that it became clear that solitary confinement is a form of torture, leaving people more anxious, antisocial, dehumanized, and aggressive than they were before.

Some of the first research on social isolation was conducted by the psychologist Harry Harlow in the 1950s. Infant rhesus monkeys were isolated in darkness for up to one year from birth. After emerging from isolation, they showed severe psychological and social disturbances, including continued self-isolation, anxiety, and inhibition. The damage was irreversible. This aptly named "pit of despair" experiment, considered unethical by today's standards, is thought to have given birth to the animal liberation movement.

When we carry the load of anxiety alone, we risk being caught, like Harlow's poor baby monkeys, in the pit of despair. Anxiety cannot be separated from our social evolution. We know deep in our DNA that one of the best ways to cope is to share the emotional load across multiple brains—our social network—whether through the simple act of holding hands or the myriad ways in which we seek and supply social support.

Anxiety is much more than the three F's: fight, flight, fear. It is the full package, protective and productive, orienting us toward rewards and binding us to the tribe. It achieves this *because* it is uncomfortable; we are hardwired to perceive and dislike that discomfort and so are driven to listen to the information anxiety provides and take the steps necessary to change a situation for the better. Anxiety contains beautiful, fractal symmetry; it evolved to give us everything we need, contained within itself, to both guide and motivate us to change situations to our advantage and to manage its intrinsic unpleasantness.

By recruiting aspects of our biology that we typically don't think of as going together—the threat, reward, and social bonding systems—it helps us handle the inherent uncertainty in the world. Anxiety, like hope, gives us the endurance to keep going and the focus and energy to work toward what we desire. When we think of them this way, we see that anxiety and hope are not opposites; they are two sides of the same coin.

3

Future Tense

Choose Your Own Adventure

Anxiety for the future time disposeth men to enquire
into the causes of things.

—Thomas Hobbes, *Leviathan*

A giant spiral staircase winds up the light-filled foyer, flanked
by exquisite Tibetan lion sculptures standing guard. Paint-
ings of mandalas and statues of the Buddha are tastefully
positioned throughout the space. To the right is something
that seems quite out of place in the Rubin Museum of Art,
devoted as it is to Himalayan cultures: a huge wall, half
blue, half red, covered from floor to ceiling with hundreds
of white cards. Walking over to it, I can make out writing
on each of the cards, like secret messages in plain sight. I am
not alone in this discovery. My six-year-old daughter, Nandini, runs up to the wall, reads a few cards, looks around,
and—as usual—is the first to figure it out. "They want *us*
to make the art!"

On a nearby table are piles of cards with either "I am hopeful because . . ." or "I am anxious because . . ." written at the top. Nandini selects a card for hope and finishes the sentence "I am hopeful because . . ." with one of the words she has mastered spelling: "love." She proudly hangs the card on one of the dozens of hooks affixed to the blue half of the wall. The cards next to hers read, "I am hopeful because . . ." "No matter how lonely you are, the world lends itself to your imagination"; "People with bad GPAs can still be successful"; "She said yes!"

On the red side of the wall are scores of cards with "I am anxious because . . ." at the top, followed by sentences such as "I don't know where to go next"; "Racism is destroying us"; "I don't know if I will find love again"; "My daughter is struggling"; "I despise wisdom because it gives me false hope."

My nine-year-old son, Kavi, has been studying the patchwork quilt of cards the whole time. He points out an interesting pattern: the cards for anxiety are often identical to the ones for hope: "I'm anxious because I have a job interview"; "I'm hopeful because I have a job interview"; "I'm anxious because people are fighting over politics"; "I'm hopeful because people are fighting over politics."

He asks me, "How can we be anxious and hopeful about the same thing?"

Here, at *A Monument for the Anxious and Hopeful*, we visitors experience how closely anxiety and hope are intertwined, how they ebb and flow like a wave, sometimes playing off each other, sometimes echoing or contradicting each other, always moving together to nudge us toward our imagined futures. As described by the monument's creators, "Anxiety and hope are defined by a moment that has yet to arrive."

In other words, anxiety and hope make us into mental time travelers, heading straight into the future.

Anxiety has shaped the course of human history. To understand how, we must first explore the radical changes in the human species that allow us to be anxious and take a tour of the varieties of future thinking that determine how well we live with, and what we achieve with, our anxiety.

My Mind on My Future and My Future on My Mind

It was only a couple million years ago, a tiny blip on the screen of our evolutionary history, that we *Homo sapiens* diverged from our ancestors *Homo habilis* and *Homo erectus* in one particular way: we developed a big brain. How big? Almost three times the size of what had filled those ancestors' skulls. Yet our entire brain didn't balloon in size, just one very special part: the prefrontal cortex. This is the area that helps us control our emotions and behaviors. That function alone would justify the increased energy required to support our bigger brain. But the prefrontal cortex also enables humans to do something else that no other animals can: to pierce the boundary between thought and reality by imagining things that aren't happening. In other words, thanks to the prefrontal cortex, human brains are reality simulators. We can experience things in our heads before we try them out in real life: we can imagine events that have not happened, relive moments of the past, and visualize possible outcomes of experiences before they take place.

The ability to simulate reality and to imagine ourselves backward and forward in time is right up there with the opposable thumb as an evolutionary advantage, enabling us to go from cave dwellers to civilization builders. When we can rehearse our actions, we can imagine what might go wrong, make better decisions, and figure out how to strive toward the future we want and need.

We use mental simulation all the time, from the smallest decision to the highest level of challenge. Does it seem like a good idea to tell a few jokes before we inform our boss about our company's projected profit losses—just to lighten the mood? We don't need to try it out first to know that it is probably a terrible idea. Elite athletes, from prima ballerinas to Olympians, mentally simulate performance and competition as an essential part of their training. The Olympic champion Michael Phelps visualized the details of every upcoming race—exactly what he would need to do for every dive, stroke, flip, and glide as well as potential problems, from fogged-up glasses to disqualification—every night and morning. Whether the best- or worst-case scenario came to pass, he had visualized and prepared for it, ready for anything.

Thank you, prefrontal cortex.

Varieties of Future Thinking

Although many people argue that the key to happiness is to *be here now*—no simulation required—our remarkable ability to imagine the future, provided by the prefrontal cortex, confers immense advantages. Anxiety, which motivates and energizes us to care about what lies ahead, can help make us ready for anything. Yet it is the rich variety of human thinking about the future that determines what we do with our anxiety—whether we take advantage of it or it takes advantage of us.

The ways we think about the future tend to fall into grooves ranging from optimism to pessimism and from believing that we are in control to feeling that we are captives of fate. Anxiety falls along these grooves in some surprising ways.

We all know optimism—the assumption that the future is likely to be favorable and that we will achieve and succeed more than we will fall short. Most of us tend to be optimistic. Ask a young adult, as dozens of studies have done, "Compared to other people of your age, gender, and background, how likely are you to win an award recognizing your achievements, get a high-paying job, marry the love of your life, and live past eighty? How likely are you to develop a drinking problem, get fired, contract a venereal disease, get divorced, or die of lung cancer?" The majority of people think that the likelihood of the good outcomes occurring is significantly above average and of the bad, below average—despite the fact that the statistically correct answer is that our chances are just that: chance.

Optimism has clear real-life benefits, from increasing our motivation and supercharging the efforts we make to pursue our goals to increasing a sense of well-being. Yet envisioning a positive future does not *necessarily* make us happier and better adjusted. We can even be optimistic in ways that can cause more harm than good.

One example of this is called *positive indulging*, a form of fantasizing in which we imagine a desired future but don't connect it to our present reality. We think, *I want to have a satisfying, high-paying job*, but then fail to remember that we don't have a degree or useful skills and prefer to work only twenty hours a week. We never picture how to get from here to there. When we fantasize in this dreamy, gauzy way, we are less likely to commit to our future goals and have fewer plans for how to overcome the obstacles. We indulge in this type of fantasizing because it feels good in the moment—research shows that it even tends to boost our short-term positive mood—but in the long run, we are likely to fall short and dwell on the emotional pain of failure.

Since it feels good to be optimistic, we often assume that pessimism is an unhealthy way to think of the future. We believe this dark side of future thinking will make us feel more anxious and depressed and will cause us to fail in achieving our goals. The truth is more complex than that.

Pessimism can lead to bad outcomes as well as good. The negative aspects of pessimism are quite clear when we look at its extremes, which often accompany anxiety disorders, such as:

CATASTROPHIZING: "It's going to be a complete, annihilating disaster."

METAWORRY, OR WORRY ABOUT WORRY: "If I become anxious and worry too much, it will harm me or cause something bad to happen."

INTOLERANCE OF UNCERTAINTY: "It is terrifying and unacceptable that my future is unknowable and unpredictable; negative events could happen at any time."

We can see these patterns of pessimism across a range of anxiety disorders. Catastrophic thinking, for example, is common among trauma survivors, who are often caught in a distressing cycle of imagining the future in reference to their past experiences ("When I look into the mirror tomorrow morning, I will see the scar from the assault") or in generally disastrous ways ("I will so completely fall apart when I interview for the new job tomorrow that they will have security escort me out"). Then there are those who suffer from metaworry, such as people diagnosed with generalized anxiety disorder (GAD). Although they chronically worry in the hope of anticipating threats and problems in order to find solutions, they also perceive that worry itself is a danger and have thoughts such as "Worrying will make me lose my mind"; "I am damaging my body with worry"; "Worry can cause a heart attack."

If pessimism becomes a habit, it can lead to the truly disruptive *pessimistic certainty*, wherein we assume not only that bad things will happen but that we will be powerless to fix them. Pessimistic certainty can exacerbate high levels of anxiety, but when it extends to the other side of the coin—the certainty that *good things won't happen*—it primes depression and suicidal thinking. When we can no longer see a potential for improvement, we might start to feel that life is not worth living.

Yet thinking about future negatives can be helpful, too. Research on aging and illness has shown that focusing on one of the biggest future negatives—our own mortality—helps us savor the present. Perceiving that our lifetimes are limited, whether because we are old or ill, causes us to prioritize healthy goals, such as making strong emotional connections with friends and family or enjoying pleasurable activities. Thinking about our inevitable death in the future drives us to pursue joy in the present.

Where does anxiety fall on this spectrum of optimism to pessimism? Surprisingly, it tends to sit right in the middle, because it's not simply about positive or negative futures; it forces us to deal with uncertainty.

Imagine that someone asks you to do the following every day for two weeks straight: "Please try to envision, in the most precise way, four negative events that could reasonably happen to you tomorrow. You can imagine anything, from everyday hassles to very serious events. Examples could be 'My hairdresser will ruin my hair while I'm already in a hurry for Julie's wedding' or 'When I take a shower in the morning, the water will suddenly turn very cold' or 'My doctor just got back test results showing that my eyesight problem is due to a tumor.'"

But what if you are instructed, "Please try to imagine, in the most precise way, four neutral, routine events that could happen tomorrow, things that you barely notice, such as brushing your teeth, taking a shower, tying your shoelaces, taking a bus, or turning on your computer."

A study did just that with about a hundred people. When they were asked to imagine two weeks of negative events, nothing much happened to their mood; their anxiety didn't increase, nor did their happiness decrease. But when they were asked to imagine neutral, humdrum, routine occurrences, their anxiety *decreased*.

That unexpected finding shows us that uncertainty—more than pessimism or optimism—is what makes anxiety unpleasant. That's because anxiety and uncertainty are so closely linked that even thinking about or planning for the most mundane, forgettable, yet predictable future events—things as simple as brushing our teeth—manages our anxious feelings. Devoted list makers—present company included—already know this.

If an evolutionary function of anxiety is to focus us on the uncertain future and motivate us to do something about it, then we also possess one more useful aspect of future thinking: we must believe that we have the power, the control, to shape the future.

When we think about the future, do we believe that we are the narrator of our own story or that we are the helpless victim of fate? Those are the extremes of the spectrum of control beliefs we all inhabit. Where we fall on that scale at any given moment has a strong impact on our emotional well-being. When we lose belief in our ability to control fate, we may seem realistic, but we will also feel more depressed. In psychology this is called *depressive realism*: being sadder but, arguably, wiser. It's a high price to pay.

Luckily, and despite evidence to the contrary, most of us err on the side of believing that we *can* control the future— even when we rationally know that it's impossible to do so. It might seem judgmental to call this magical thinking, but that is just what it is. Dozens of studies examining the variety of ways we think we can control the uncontrollable show that most of us believe that if we spin the wheel in a certain way or blow on the dice, luck will lead us to victory. One of the earliest studies in the field demonstrated that the overwhelming majority of us believe, deep in our bones, that if we pick our lottery ticket rather than being randomly given one, we have a better chance of winning. The same illusions of control apply to situations that don't rely on mere chance; we're certain that through sheer force of will, we can turn our dreams into reality while avoiding disaster.

That's because we find it natural to take credit for our successes and blame our failures on external factors. This habit of interpreting events, which psychologists call an *internal-stable-global attributional style*, assumes that we control the positive events in our lives. It's internal-stable-global because we attribute good events to our own rather than others' efforts (internal rather than external), believe that this will almost always be the case (stable rather than unstable), and are confident that it will be the same in every situation in our lives (global rather than specific). These attributions extend into the future; they are what we can expect from tomorrow and from the many tomorrows after that. Think of it as uncertainty managed: over and over again, studies have shown that what is essentially an error in thinking goes along with a healthy emotional life.

Conversely, when we reject such illusions of control of positive events, we are more likely to be depressed. Depression even turns this healthy attributional style inside out, so that we now believe that positive events are due to external, unstable, and specific causes—which means that good things happen by chance, outside our control, and only sometimes. It's hard to look forward to such a future.

Anxiety, in stark contrast to depression, embraces and leverages the internal-stable-global attributional style. When we're anxious, even intensely so, we still believe that we can make good things happen in our lives. And the most common mental action we take that helps anxiety achieve this is something we're all familiar with.

It's worry.

Worry Is the Belief That We Can Control the Future

Most of us, including myself, use the words *worry* and *anxiety* interchangeably in day-to-day life. But in psychology, they're considered to be two distinctly different things. Anxiety is a mixture of physical sensations, behaviors, and thoughts. The feelings—the butterflies in our stomach, the tightening in our throat, the overall sense of being energized and agitated—are in our bodies. The behaviors are what we do when our threat response is triggered—either fighting, fleeing, or freezing.

Our thoughts, meanwhile, are trying to figure out *why* we're anxious and what we should do about it. That thinking part is what we call worry. It has an "aboutness" that anxiety doesn't always possess. Anxiety can be free-floating, without any apparent object or focus. I feel anxious but I'm not sure why, which is distressing, so I try to soothe myself—say, with some deep breathing or, perhaps less wisely, a glass of wine. Worry, in contrast, is sharp and directed: I'm worried that I can't pay the rent. I'm worried that I'll die of the same illness as my grandfather did. When we worry, we might still reach for a glass of wine, hoping that it will help, but we also are primed to do something truly useful, such as asking: What should I do now?

By worrying, we begin to figure out how to handle anxiety-provoking situations. I need to figure out how to get more money so I can pay the rent. I have to see the doctor and get tested so I know whether or not I have a disease. Worry is agitated, persistent, and relentless, because its aim, its sole purpose, is to help us figure out how to deal with threats and make things turn out right in the end.

You can be anxious *without* worrying, as when your anxious feelings are diffuse and vague, difficult to nail down, but you can't worry without being anxious. To study worry, researchers actually instruct people how to worry, asking them to invoke specific thoughts and ideas. The feeling parts of anxiety tend to follow. The directions are:

TAKE A MOMENT TO FOCUS ON HOW YOUR BODY FEELS: your breathing, your heart rate, muscles (tune in to your shoulders and face muscles), and how you're sitting or standing (tight or relaxed). Next focus on your thoughts: What is going through your mind right now?

NOW LIST THREE THINGS THAT MAKE YOU ANXIOUS AND PICK THE ONE THAT IS THE MOST INTENSE. SPEND A WHOLE MINUTE THINKING ONLY ABOUT THAT MOST INTENSE TRIGGER OF ANXIETY. REALLY LEAN INTO IT. IF POSSIBLE, THINK ABOUT IT AS VIVIDLY AS POSSIBLE: images, details, the worst that could happen, and what you're going to do about it.

AFTER THE MINUTE IS OVER, TUNE BACK IN TO YOUR BODY. IS YOUR HEART BEATING A LITTLE FASTER, DO YOU FEEL WEAKNESS OR HEAT, STIFFNESS OR A DRY THROAT? IS YOUR BREATHING FASTER OR MORE RAPID, OR ARE THERE BUTTERFLIES IN YOUR STOMACH?

Clearly, worry doesn't feel so great. It can easily make anxiety worse, training our minds on troubles and uncertainty, triggering our fight/flight bodily responses. Since worrying doesn't feel good, why do we keep doing it? Because there is one particular aspect of it that feels extremely positive: worrying makes us feel as though we're *doing something*. When we're anxious, worry is often triggered to speed us into a mental simulation of the future, pushing us to plan what to do next. And because it feels good to believe we can control the future, we keep worrying.

I have a firsthand understanding of the thinking-planning-controlling nature of worry because of the most anxiety-provoking experiences of my life: learning about my son's congenital heart condition.

When I was expecting my first child, Kavi, we discovered that he had a serious condition that would require him to go through open-heart surgery within months of his birth. It may seem obvious where worry fits into this equation. Less obvious is the fact that worry was one of my best—albeit most exhausting—friends during the year from diagnosis to my baby's surgery and recovery before he had even reached six months of age. The terror I felt wasn't particularly helpful. My free-floating anxiety was slightly more useful because it energized me to keep going. But it was the worrying part of my anxiety that allowed me to stay one step ahead of the danger my son faced if we didn't get him the medical intervention he needed. Worry pushed me to figure out how to maximize the chances of a successful surgery and minimize the odds of the worst outcome imaginable.

My worries were legion. While still pregnant, I worried about his prognosis and how sick he would be once he was born. I tried over and over to imagine what it would be like to care for a sick baby; I wanted to be like an Olympic swimmer but of motherhood, picturing every stroke of the race that would be my son's medical emergency. I snapped into information-gathering mode: I read every paper ever published on Kavi's condition, scoured the congenital heart disease community blog sites, and asked our nurses and doctors a million questions at our weekly prenatal checkups tracking his progress through ultrasounds and echocardiograms.

Worry helped me plan. Having his operation when he was several months old rather than immediately after birth would give his heart time to grow bigger and stronger, so we hired a baby nurse to extend the time we could care for him at home. I worried about finding the best surgeon. We found some outstanding ones and had to choose among them—should we go with the one who had a better bedside manner or the one who everyone told us would keep a laser focus and a steady hand even if a bomb went off in the next room? (We went with the steady hand.) Every week, I imagined best- and worst-case scenarios, spoke with specialists about every contingency, and tried, as much as possible, to plan every single detail of his care. And of course, I worried: How in the world will we get through this?

In the end, it was the worry that helped us get through. While it allowed me to prepare at a highly effective level, it also helped me survive emotionally, because I never stopped believing that if I planned and worked and thought hard enough, our son would live and thrive—even though I also knew that total control over the future is an illusion. My worry was my belief that we could fight for our son's survival in the face of a disease that would have been a death sentence not so long ago.

Don't get me wrong; worry isn't always helpful. When chronic and extreme, it disrupts rather than assists our ability to create the future we want. For example, worry is a key component of the most common of the anxiety disorders, generalized anxiety disorder (GAD). In earlier centuries, GAD was termed *pantophobia*, or fear of everything. That makes sense, because someone diagnosed with it worries indiscriminately—about world events, finances, health, appearance, family, friends, school, work. This makes worry extremely time consuming. Generalized worry is also distressing because it feels out of control and persistent, like a perpetual motion machine in our mind. Such a juggernaut feels frightening, as though it could lead to a mental or physical breakdown.

Penn State researchers illustrated this peril in a 2004 study. They asked people diagnosed with GAD to do two completely opposite things: first, worry about something that really bothered them, and second, calmly focus all their attention on their breathing in order to relax. During the breathing exercise, they wrote down whether they still felt distracted by lingering worries. It turned out that even then they were plagued by intrusive worries, an inability to focus, and feelings of restlessness, tension, and fatigue. In other words, they couldn't turn their worry off. In its most extreme form, it becomes so automatic that we worry even during times of safety and relaxation.

Choose Your Own Adventure

Future thinking can help or it can get in the way, but launching our minds ahead toward the moment that has yet to arrive always has a certain emotional quality; maybe we feel the *frisson* of uncertainty, the heightened focus and quickened heartbeat, or a little burst of adrenaline as we marshal our resources to prepare for the unknown. Our minds have entered the future tense, where uncertainty, anxiety, and hope all live together.

This simulation energizes us by its very nature. The past and present tenses can't give us this edge or urgency. Anxiety tells us that waiting around for the future to happen could be bad, so we had better create the outcome we want. It's not unlike a *Choose Your Own Adventure* book.

Once my son's surgery date was set, my brain went to work, planning: We'll hire a car service and leave for the hospital at 6:00 a.m. That will get us there in plenty of time, and we won't have to worry about driving. Once we check in, we will meet with a nurse, and I'll be able to ask any final questions I have. In fact, I'll write down any lingering questions the night before, just in case my mind goes blank when we arrive. After the meeting with the nurse, the anesthesiologist will explain the procedure and give Kavi a sedative that will put him to sleep. That will be a relief; I won't have to think about whether he's scared. I wonder if I should be the one to carry Kavi to the operating room. Will it feel better or worse if his mother hands him over to the surgical team?

If I carry him, go to page 20. If my husband does, skip to page 53.

Perhaps dark humor doesn't suit this situation. But imagining what will happen next and then choosing among the possible paths to take is seeing my son's surgery in the future tense. Worries and planning enter the foreground of my mind. My thoughts move along quickly. Even though it's only a mental simulation, my heart beats faster as if in preparation for the event. I feel pangs of anxiety, hope, dread, confusion, and other things as I face the uncertain future, but I also feel more focused. By the time I return to the present moment—I give myself a break so I don't go too far down the rabbit hole of worry—I have not forgotten the danger we will face, but I feel doubly prepared to do everything I can to ensure a good outcome.

I experience the surgery very differently from the perspective of the past and present tenses. In the present tense, I rode an ever-moving torrent of thoughts and perceptions, feelings, and ideas—some about the surgery and some about other things: "Oh, no, handing him over to the surgical team myself is a terrible idea. I am not sure I can let go of him. And this huge, bright room full of gleaming metal instruments—I'm going to either faint or throw up. I better not throw up on the doctor! Okay, disaster averted, he's in the operating room. I just have to remember that it will be okay. Our surgeon is the best. This kind of surgery is a walk in the park for him. Now I just have to make my way to the waiting room. Okay, here's the waiting room. It's really, really quiet here. People are murmuring in the corner.

Where's my husband? Oh, there he is. I feel so grateful that our friends and family are here with us. Ugh, this coffee tastes awful and is making me more nauseous. Why can't I stop drinking it? How much time has gone by? One hour? Three hours? Is that the surgeon opening the door, is the surgery done? No. Is that him? No. Now? No. When will it be over? Why is someone wearing such strong perfume?" My mind gallops.

In the past tense, however, time slows down and expands as I tell and retell the story of the surgery to myself. In one version of the story, I focus on my parade of feelings: the cold fear I felt while I waited; the painful imaginings going through my mind of the doctor cutting into Kavi's chest, cracking his rib cage open, and stopping his tiny heart so they can operate on it; my growing exhaustion as the hours ticked by; and thankfully, the inexpressible relief when the surgeon finally came out to tell us that the surgery had gone perfectly. In another version of the past, there were idiosyncratic details and images that defined the experience: the antiseptic look of the waiting room; the visit from the anesthesiologist, who recommended—I kid you not—going to a great sandwich place on the corner while we waited—as if we could eat!; the moment the door opened and it wasn't the surgeon as we had hoped, but it was one of our dear friends, and we felt a surge of comfort; and after the surgery, when Kavi was recovering in the hospital and doing so well, the moment of certainty that he not only would be fine but would thrive. The more I elaborate on and retell the positive details of the past, going over the same moments, adding details and interpretations here and there, the better I feel. I like to sit and immerse myself in that story of the past, like a warm bath.

The past tense is slow and narrative, giving us the ability to create a comfortable story to tell. The present tense is a circuitous stream of experiences, meandering along. But the future tense is dynamic, full of momentum, surging forward toward an ending that hasn't happened yet but that we want to make happen.

Paradise Lost

The sine qua non of anxiety is the future tense. When we're anxious, the question "What will happen next?" contains both positivity and peril. It's as if the future is a weak radio signal. As we twirl the tuning knob, trying to find the right setting, anxiety nudges us to tune in to the channel representing the future we want. Indeed, our marvelous human brains—our reality simulators—evolved not to stumble into the future but to imagine it so we can create it.

That's why if we want to sit back and relax, the future tense is probably not the best choice. As we'll see in chapter 10, that's where the present tense rules. But if, instead, we want to get things done and plan ahead for things that matter to us, we can choose no better than the future—albeit in the right doses. This is what makes anxiety both protective and productive—and makes it a prime force driving human achievement and ingenuity. At the beginning of this chapter, my son Kavi asked, "How can we be anxious and hopeful about the same thing?" The answer I gave him was "We're only anxious when we care. And there is so much to care about."

Ironically, as we'll see in the next section, some of our greatest achievements—language, philosophy, religion, and science—have steadily eroded our ability to use anxiety to pursue the things we care about. Our current beliefs about anxiety have almost succeeded in turning it from an advantage into a disadvantage. Almost.

PART II

How We Were Misled About Anxiety

4

The Anxiety-as-Disease Story

As we've seen, anxiety is not just a blip on our emotional screen; humans are *made* to be anxious. Anxiety is embedded in our ancient defensive biology and intrinsically connected to our deep-seated need for human connection. Anxiety is what makes us different from other animals. Without it, we might never have become civilization builders—or even survived as a species.

But we seem to have squandered our relationship with anxiety. When we look at ourselves today in the twenty-first century, we see that we treat even the mildest of anxious feelings like an unwanted burden. We dread anxiety so much that we'll do just about anything to evade or suppress it.

We treat it like a sickness.

The transformation of anxiety from advantageous emotion into unwanted disease did not happen overnight. It took a thousand years to deceive ourselves into believing that this evolutionary triumph is an illness leading us down the twisted path toward madness and terror. To tell the story, we have to start at the roots of modern-day medical sciences in the Dark Ages.

Going Medieval on Anxiety

In the early medieval period in western Europe, the Roman Empire was in its final stages of collapse and the Catholic Church had taken center stage in people's lives, shaping everything from how they worshipped, what they ate, and when they worked to how they thought of life, death, and the afterlife.

At the time, the word *anxiety* didn't mean anything remotely like what it does today. Then people thought of it as a bodily sensation encapsulated by the word's etymological roots—the Latin *angere*, "to choke," and the even older Proto-Indo-European *angh*, "painfully constricted." Also unlike today—when we freely use the word to describe any feeling of distress or worry—the medieval terms for anxiety were almost never part of common conversation—*anxietas* in Latin, *anguish* in English, *anguisse* in French, and *angst* in Germanic and Scandinavian languages.

The Church, however, changed all that by making anxiety a key component of spiritual life. *Anxiety* became the go-to word to describe the anguished suffering of the soul, entrapped by sin, passionately yearning for redemption, and terrified of the eternal tortures of Hell, captured in such exquisite detail in Dante Alighieri's fourteenth-century epic poem, *The Divine Comedy*.

Indeed, the opening lines of *Inferno*, the first book of *The Divine Comedy*, begin by evoking otherworldly anxiety as the protagonist, Dante the Pilgrim, lost in a dark wood, begins his terrifying journey through the nine circles of Hell and Purgatory on his way to Heaven:

> *Midway upon the journey of our life*
> *I found myself within a forest dark,*
> *For the straightforward pathway had been lost.*
> *Ah me! how hard a thing it is to say*
> *What was this forest savage, rough, and stern,*
> *Which in the very thought renews the fear.*
> *(Inferno, canto I, 1–6)*

Each concentric circle of Hell is a city structured around specific tortures, like a planned urban space, and holding worse and worse sinners the deeper Dante goes—from lakes of fire and fiery sands to crucifixions, burial in open graves, and submersion in bile. Written in vernacular Italian and illustrated with startling imagery, *Inferno* describes in everyday language the eternal agonies sinners will suffer in the afterlife. With the horrors of Hell and the threat of damnation now dominating the medieval mind, anxiety became a familiar companion. It joined the ranks of other pivotal Sunday sermon abstractions such as hope, faith, conscience, purity, and salvation.

As the significance of anxiety became more spiritual, its treatment also changed. Now the healers of the soul, Catholic priests, prescribed and administered the interventions of confession, penance, and prayer. As Saint Augustine taught, "God can relieve your troubles only if you in your anxiety cling to Him."

This concept of anxiety as a spiritual condition requiring divine relief became common throughout the Holy Roman Empire, which spanned what are today forty-eight countries as far north as Scotland, down through all of Europe, and into Asia and northern Africa. Yet soon another paradigm shift would push anxiety further along on its journey.

Enlightened or Not, Here I Come

In the seventeenth century, notions of freedom and individualism pushed people to question old ways and old authorities. *Sapere aude*—Dare to know—was the motto of the Enlightenment. Thinkers and scientists defied the strictures of the Church—and were often burned at the stake for it. They used the tools of empiricism, scientific observation, and mathematics to explain the mysteries of the natural world and achieve new technological feats.

One of the most important books of the era was *The Anatomy of Melancholy*, written in 1621 by Robert Burton, a university scholar and sometime librarian. Although he presented it as a medical textbook, the encyclopedic overview of pathologies of emotion was equal parts science, philosophy, and literature. While consisting of quotations from ancient medical authorities such as Hippocrates and Galen, it was also chock full of empirical observations, case studies, and sympathetic portrayals of emotional distress. Melancholia was not limited to depression but encompassed anxiety and a range of bodily complaints, hallucinations, and delusions. Burton even included in his list religious melancholia, or the defects in religious feeling experienced by "atheists, epicures, and infidels."

Burton's goal was to deconstruct and dissect melancholia first in terms of causes and symptoms and then in terms of cures—as one would for any other disease. His observations aren't so different from our modern views of anxiety disorders, which cause sufferers to be plagued with worries, sick with anxiety, simmered until the "foul fiend of fear" caused them to turn "red, pale, tremble, sweat, it makes sudden cold and heat come over the body, palpitation of the heart, syncope, &c." He described people becoming "amazed and astonished with fear."

Burton was an unlikely candidate for "the first systematic psychiatrist," as the late French American historian Jacques Barzun called him. His education at Oxford was unusually lengthy, perhaps due to a bout of melancholia. His protracted and extensive studies included nearly every science of his day, from psychology and physiology, to astronomy, theology, and demonology—all of which informed *The Anatomy*.

The digressive and labyrinthine volume was reprinted no fewer than five times during his lifetime and was read by luminaries spanning centuries, including Benjamin Franklin, John Keats (who said it was his favorite book), Samuel Taylor Coleridge, O. Henry, the artist Cy Twombly, and the writer Jorge Luis Borges. Samuel Beckett and Nick Cave have both referenced it with admiration.

The Anatomy of Melancholy is a seminal work in the transformation of anxiety into a disease. But the philosophical upheavals of the seventeenth and eighteenth centuries would take it a step further, locating Burton's hellish "foul fiend of fear" in the mind rather than in the soul and arguing that irrational emotion can be controlled only by rational thinking. That was, after all, the Age of Reason, when faith in the Church's explanations waned.

Yet the new post-Enlightenment mind—capable of thinking, imagining the future, and constructing reality—was also a vulnerable mind, robbed of the medieval certainty of faith. Anxiety emerged at those fault lines, where free will collided with the vicissitudes of random fate and unpredictable passions. Later generations would call it *existential angst*.

Indeed, those living through that paradigm shift often paid the price in the currency of anxiety. England in the eighteenth century was the most liberal, progressive, and modern society in the world. Yet it was also one in which anxiety and mental health problems seemed ubiquitous. Suicide rates skyrocketed during that period, so much so that suicide was termed "the English disease." It was as if that free, unfettered society had, as François-René de Chateaubriand wrote at the end of the eighteenth century, grown "sick with anxiety and indecision."

To many in the Western world, the fact that the mind was free but separate from the heavenly soul was undeniable—but also unbearable. There was now a need for new, modern healers of the soul. Early psychologists and psychiatrists—called *alienists* and *mentalists*—heeded the call. They would solidify—and make inevitable—the anxiety-as-disease story.

The Medicalization of Anxiety: From Phrenology to "Rat Man"

As the nineteenth century dawned, the medical community was greatly concerned with the treatment of what was then understood to be diseases of the mind—mental illness. Pseudoscientific theories such as phrenology—which used the analysis of bumps on the skull to predict emotional and personality traits—jump-started the "somatogenic" versus "psychogenic" debate. The former argued that mental illness originated in the brain and body—the Latin *soma*—like any other disease. The psychogenic side countered that mental illness must have its origins in psychological states and experiences such as trauma. In the latter part of the nineteenth century, Sigmund Freud was the best known and most influential of the psychogenic theorists—despite the fact that as a medical doctor by training, he started in the somatogenic camp, believing that anxiety and mental illness were purely biological phenomena.

Whether psychological or biological in origin, anxiety was a prime focus of the growing movement to treat mental illness with standardized therapies and medications. That represented an improvement; before, bouts of anxiety had been thought to be due to "the vapors" and treated with smelling salts or even exorcism.

Hysteria was among the most common anxiety diagnoses in the nineteenth century. Derived from the Greek word for "uterus," hysteria was considered a female problem resulting from a "wandering uterus" moving willy-nilly around the body and blocking the healthy circulation of "humors." Overly emotional and irrationally upset, a hysterical person experienced a strange mix of symptoms as diverse as shortness of breath, fainting, paralysis, pain, deafness, and hallucinations. Despite growing medical knowledge about the implausibility of a peripatetic uterus, Freud and his followers often treated hysteria. They did so, however, with relative scientific rigor—using talk therapy to target the suppressed memories and desires they believed gave rise to hysteria.

Despite the fact that the clinical treatment of hysteria and other forms of anxiety was becoming more common and accepted, English-language textbooks of psychology and psychiatry didn't use the word until the 1930s and only after the translation of Freud's 1926 tract *Hemmung, Symptom und Angst* into English as *The Problem of Anxiety* in 1936. Interestingly, Freud, like his German-speaking compatriots, used the term *Angst*, a word known from their childhood. Yet the word *anxiety* trickled down into the English-speaking consciousness. In 1947, after the catastrophic losses and horrors of the two world wars, W. H. Auden gave a name to the distress of his times in the epic poem *The Age of Anxiety*.

Freud—and many therapists who came after—believed that anxiety was a common and mostly healthy emotion. Yet as Freudian theories of mental illness hinged more and more on the role of trauma, repression, and neuroses—all of which triggered anxiety—it became central to the psychiatric endeavor. Mental illness almost couldn't be conceived of without anxiety.

Take one of the most famous of Freud's case studies, "Little Hans." The patient, whose real name was Herbert, was the son of a friend of Freud, a famous music critic of the period, Max Graf. As a young boy, Herbert witnessed a horse carrying a heavily loaded cart collapse and die in the street. Following that traumatic event, the five-year-old developed a fear of horses, refused to leave the house for fear of seeing one, and was tormented by the thought that a horse would come inside and bite him as punishment for wishing that the horse would lie down and die.

In his report on the case, published in his 1909 paper "Analysis of a Phobia in a Five-Year-Old Boy," Freud argued that the boy's fear of horses was not directly caused by his witnessing the animal's collapse in the street. Rather, it was a displacement of fear of his father onto the animals, whose blinkers made them resemble a man wearing glasses, which his father did. The boy unconsciously wished that his father would go away or die because he regarded him as a competitor for his mother's love—the so-called Oedipus complex. That caused Herbert anxiety, including the fear that he would be castrated by his father, which could be resolved only through the defense mechanism of displacement—i.e., transferring his fears of his father onto horses. Because his animosity toward his father, whom he loved, was intolerable, treatment sought to help him express his anxieties in order to relieve them, like releasing a pressure valve. When Herbert was able to describe his fantasies, his fear of horses disappeared, indicating the resolution of his castration anxiety and an acceptance of his love for his mother.

Another famous Freudian patient was called "Rat Man"; his obsessions were described in Freud's 1909 paper "Notes upon a Case of Obsessional Neurosis." The patient suffered for years from obsessive worries that misfortune would affect a relative or close friend unless he carried out specific compulsive behaviors. Even after the death of his father, he continued to be plagued by the worry that harm would befall him. Rat Man's symptoms look a lot like what we would term obsessive-compulsive disorder, or OCD, today.

Freud used techniques such as free association to uncover the repressed memories that he believed drove those obsessive worries. A key memory was from his military service, when Rat Man had learned the terrible details of a torture method in which a person was placed in a container of live rats. The creatures would have to chew through the victim to escape. The image stayed with poor Rat Man, and it was the torture that he feared would befall his relatives or friends. He also believed that if he could pay someone to collect parcels from the post office for him, it would somehow prevent that terrible fate. He would become increasingly anxious until someone helped him complete the magically effective ritual.

What did Freud make of Rat Man's obsessions? He believed that they resulted from a completely different repressed anxiety—his repressed childhood fear that his father would severely punish him if he found out that Rat Man had had early sexual experiences with his childhood governess. When his fear of punishment was repressed, the hostility he felt toward his father was also banished to his subconscious. What did Rat Man do with that mishmash of repressed anxieties and hostility? He replaced it with the fear of bizarre misfortune that would kill his father and later, all his loved ones. It took Freud eleven months to bring all those anxieties from the darkness of the subconscious into the light of consciousness, but once he did, Rat Man was reportedly cured of his obsessions.

Those classic—and peculiar—Freudian case studies make clear that anxiety was the foundation of psychoanalytic theory, which dominated psychology and psychiatry throughout the early decades of the field's development. Anxiety was a linchpin of mental illness. It was dangerous.

Yet before anxiety achieved its final apotheosis, being transformed into a disease, it needed to be medicalized. That was achieved through the *Diagnostic and Statistical Manual of Mental Disorders*, or *DSM*.

The *DSM* defines the landscape of mental health and illness. It is the system through which mental illnesses are diagnosed, using categories that distinguish among different types of anxiety disorders and separate them from other mental illnesses, such as major depressive disorder and psychosis. The very first *DSM* was published in the early 1950s. Extensively revised over the decades to its current fifth edition, it has changed in thousands of ways. But an overarching trend has determined how we conceive of the disease of anxiety. In 1980, the *DSM* went from focusing on theoretical dimensions of anxiety—calling any problem involving anxiety an *anxiety neurosis*—to categorizing and defining distinct types of anxiety-related diseases, with a checklist of criteria to diagnose each one. For example:

Do you have marked fear or anxiety about two (or more) of the following five situations?

1. Using public transportation, such as automobiles, buses, trains, ships, or planes
2. Being in open spaces, such as parking lots, marketplaces, or bridges
3. Being in enclosed places, such as shops, theaters, or cinemas
4. Standing in line or being in a crowd
5. Being outside the home alone

If so, and if you avoid and disproportionately fear these situations and these patterns persist for six months or more, then according to the *DSM*, you have agoraphobia, the fear of public places. There's no wondering; you *have* it, with medical certainty, and it should be treated as such—with specific therapies and medicines.

Used mainly in the United States, the *DSM* has been widely adopted by clinicians, researchers, regulatory agencies, pharmaceutical companies, legal professionals, insurance companies, and beyond. It's everywhere. This isn't to say that the *DSM* is bad or unhelpful. Diagnosing a problem that is causing great disruptions, that is a source of human suffering, is an effective way to develop solutions. But the *DSM* succeeded in making the anxiety-as-disease story so complete, so systematic, that it dominates our concept of anxiety today. By medicalizing anxiety, we reckon that we have made it understandable and manageable. We have forgotten that it's not always a disease.

The Danger of Safe Spaces

Another consequence of the anxiety-as-disease story is the notion of "safe spaces."

A safe space is a literal or metaphorical place for people to come together without being subjected to bias, conflict, criticism, or threat. Some of the earliest safe spaces can be traced back to feminists and gay activists in the 1960s, when safe spaces where the places for those marginalized groups to come together without fear of prejudice or ridicule.

Today, safe spaces are often found on college campuses, but the very first safe spaces emerged in post–World War II corporate America, created by one of the fathers of social psychology, Kurt Lewin. As the director of the Research Center for Group Dynamics at MIT in the 1940s, Lewin was well known as an expert in small-group interactions— it's because of Lewin that we use the term *social dynamics* and we give "feedback" to our colleagues. But he was also an early advocate of "action research," in which theory is put into action in pursuit of social justice. In 1946, he received a call from the director of the Connecticut Inter-racial Commission, who wanted to find effective ways to combat religious and racial prejudices. In response, his first workshops, conducted as leadership training programs for corporate executives, laid the foundations for what we call *sensitivity training* today.

The core assumption of sensitivity training, inspired by psychotherapy, was that change in a social group, such as a workplace, could happen only when people honestly challenged one another in small groups without judgment. To create such psychological safe spaces, participants in sensitivity training had to agree to speak honestly, maintain confidentiality, and suspend judgment. Only then could they discuss one another's implicit biases and unhelpful behaviors and point out the ways in which they compromised their leadership ability, harmed others, and disrupted the organization.

The topic of sensitivity training could be anything, but given its origins in concerns about religious and racial prejudice, those biases were often the focus of discussion. The point of the safe space was to free people to share their true thoughts and feelings without fear of condemnation while still understanding that they were hoping to change. So when a White executive admits that she feels intimidated by her Black male employee because of his race, or a Black executive admits that he feels angry toward an Asian-American colleague because he perceives her as benefiting from nepotism in ways he has not, they share these ideas believing that they will not be maligned as racists. The goal is to give and receive honest and sometimes difficult feedback in order to change.

The notion of safe spaces has itself radically changed in the twenty-first century. Now safe spaces are prohibited from being emotionally raw because they are meant to protect people's feelings—against racism, sexism, bias, and hate speech, as well as against opinions, debate, and conflict that some find distressing.

A 2015 *New York Times* opinion piece might have been the first place some of us heard of safe spaces. In it, Judith Shulevitz described a controversy at Brown University over the scheduled debate between the feminist authors Wendy McElroy and Jessica Valenti on the concept of rape culture. Valenti is a proponent of the idea that prevailing social attitudes in the United States normalize and trivialize sexual assault and abuse, whereas McElroy disagrees. Some Brown students argued that McElroy should not be brought in as a speaker because, whether or not they attended the debate, her viewpoint would cause emotional harm to students, particularly those who were survivors of sexual assault, as well as those who were disturbed by her viewpoint.

Though the effort to disinvite McElroy from the debate failed, the president of Brown, Christina Paxson, responded to students' concerns by organizing an additional talk on rape culture—without debate—and created a safe space where those who felt triggered by the topic could rest and recover. There were calming music, cookies, pillows, and blankets, as well as students and staff ready to provide emotional support. Some who went to the safe space felt threatened by memories of their personal trauma, but others felt threatened by the pain of the controversy over the speakers. As one student who retreated to the safe space told the *New York Times* reporter, "I was feeling bombarded by a lot of viewpoints that really go against my dearly and closely held beliefs."

It's worthwhile noting that equating dissenting opinions with emotional harm runs counter to the original intention of safe spaces. In sensitivity training, safe spaces contain difficult conversations that are facilitated through self-restraint, suspension of judgment, honesty, and feedback. Bigotry and bias are pointed out, rather than avoided. The conversations can be difficult, especially if people are being honest with one another. Safe spaces today, in contrast, have become places where the difficult parts of conversations are deemed dangerous and so are excised because they cause distress and anxiety.

In the debate about safe spaces, there are two primary points. On the one hand, there's the issue of whether the demand for safe spaces on college campuses and the characterization of contrary opinions as emotionally harmful infantilizes students and erodes free speech. Some people argue that safe spaces contribute to echo chambers in which we're surrounded by like-minded people and insulated from ideas that challenge or contradict our own—a barrier to the democratic ideal of the free exchange of ideas.

On the other hand, there's the issue of whether certain emotionally distressing ideas really do cause psychological harm. The practice of issuing trigger warnings is pertinent here. Trigger warnings are essentially content warnings that a work contains writing, images, or ideas that may be distressing to some people, especially in the context of sexual violence and mental illness. They've been a part of internet communities for years, primarily for the benefit of people with post-traumatic stress disorder who may want to avoid anything that reminds them of their trauma.

But it's the more recent use of these warnings in the classroom that has "triggered" debate. Some academics worry that trigger warnings teach students to avoid uncomfortable ideas, thus compromising their ability to engage rationally with ideas, arguments, and views they find challenging. Yet it's for the same reason that many professors are staunch supporters of trigger warnings. They believe that the warnings provide students the opportunity to prepare themselves for a vivid reminder of a trauma or a potentially distressing topic, so that they can manage their reactions and continue learning. In other words, they feel that when a student is in the grip of a strong emotion—or a trauma-prompted flashback or panic attack—he or she can't be expected to think straight, let alone learn.

The evidence thus far, however, suggests that trigger warnings don't help when it comes to managing distress and may even do some harm. A study from 2021 provided a group of college students and internet users with trigger warnings before they viewed negative material and compared those participants to a group who received no such content warnings. Those in the two groups reported similar levels of negative emotion, intrusions, and avoidance whether or not they had received a trigger warning and whether or not they reported a history of trauma. In a 2018 study, several hundred participants were randomly assigned to either receive trigger warnings or not prior to reading literary passages that varied in potentially disturbing content. Participants in the trigger-warning group reported *greater* increases in anxiety—especially when they believed that words can cause harm. This suggests that trigger warnings may inadvertently undermine emotional resilience and cause even more distress for some.

Setting trigger warnings and keeping ourselves safe from ideas—and the anxiety they cause—seem to do little good and may even make things worse. And if being forewarned doesn't mean being forearmed, issuing a warning about the dangers of strong emotion may just serve to perpetuate the belief that difficult emotions do us harm.

The Age of Anxiety

From the medieval Church to the Age of Reason to the halls of medicine, we have been taught the anxiety-as-disease story so thoroughly that we have memorized it chapter and verse.

Each era has advanced the disease story of anxiety at the expense of thinking of it as a normal human emotion. We have been inexorably convinced by all sides that anxiety and suffering go hand in hand. Anxiety remains Burton's "foul fiend of fear."

If you doubt this, observe how we science and health professionals have turned controlling and eradicating anxiety into a cottage industry—whether through therapy, pharmaceuticals, or teaching meditation. We have conducted thousands of rigorous experimental studies deconstructing anxiety, developed gold-standard evidence-based therapies and medications to dull the emotion, and published hundreds if not thousands of self-help books about coping with it. Yet these solutions have largely failed to affect the rates of problematic, debilitating anguish. Anxiety is on the rise, and our kids may be particularly at risk. The good news is that some of them are questioning what they're told about anxiety. They know that something is off.

I learned this firsthand when I met with a group of middle school students on a sunny winter day in Manhattan. Every year, public school administrators elect a student council that goes on to identify and pursue a mission with high potential for positive impact. I had been invited to consult with the District Two student council because its chosen mission was to advocate for improvements in school mental health services.

I soon learned that those twelve- to fourteen-year-olds were exceptionally ambitious. They had divided themselves into three work groups, each with a specific goal: one group focused on convincing legislators to fund peer-to-peer counseling services, another on obtaining funding from the city council to hire more counselors for their schools, and a third on getting state representatives to propose laws that would require more mental health funding for schools across the state.

Why had they chosen such ambitious goals? Because, as they told me, they had looked around at high schoolers, just a couple years older than they, and saw how much they were struggling—most of them with anxiety but many also with depression, addiction, and self-harm. Getting services for those students was among their top concerns, but so was enlisting the support of adults and professionals *now*, in middle school, before problems really accelerated.

Yet adults were not helping in the way they had hoped. The kids had already faced a panoply of noes from them—no, we don't have the budget for that; no, that's moving too fast; and no, that's impossible. Complicating matters, even the most well-meaning adults didn't seem to have the answers. And when they saw kids grapple with anxiety, they panicked, acting as though they wanted to wipe out any sign of it like an affliction, extract anxiety like a rotten tooth. It wasn't helping.

One student put her finger on the dilemma when she said, "The adults that actually try to help us don't know what to do. They act like they can take away our anxiety. But it's a part of us, so *can* they take it away? *Should* they?"

Until we can answer no to both questions, we'll be stuck telling ourselves the wrong story of anxiety—and making the dire mistake of constantly trying to rid ourselves of it.

5

Comfortably Numb

"We live in the midst of alarms; anxiety beclouds the future; we expect some new disaster with each newspaper we read."

What better description of how we feel in the opening decades of the twenty-first century—a time of global pandemic, viral misinformation, political upheaval, economic inequality, and the threat of irreversible environmental destruction?

But they are the words of Abraham Lincoln, spoken several years before the Civil War, another fractious and devastating period of US history.

Anxiety, then as now, was the word that explained the pain of our fears and uncertainties. It even lent its name to our era—ever since W. H. Auden published his epic poem *The Age of Anxiety* in 1947, as millions of people were still reeling from the trauma of two world wars.

Perhaps that is because many of us can no longer leverage faith, the bonds of community, or the support of our institutions—the traditional bulwarks of certainty—to manage our anxieties. But manage them we must. So we've turned to the authorities we still believe in, the high priests of the modern world: the scientists and doctors. Most are driven by the most laudable of goals—to relieve pain—but when it comes to anxiety, they have let us down. Stupendously. Like all of us, medical professionals have come to believe the anxiety-as-disease story and have taken it to the next level, devising foolproof ways to rid us of our worries and angst—temporarily.

This "achievement" is in large part thanks to the miracle of modern pharmaceuticals that literally suppress even the whisper of anxious feelings. These drugs soothe and sedate us. Over the past sixty-plus years, they have taken center stage in our relationship with anxiety. Even under a cloud of controversy and debate, their sheer ubiquity has created a societywide mindset; when there's emotional pain, we take a pill to dull it. We have been convinced that the best approach to dealing with anxiety is to become comfortably numb.

A Brief History of Chemical Calm

In the first half of the twentieth century, barbiturates—sedatives and tranquilizers—were the go-to medicine to suppress anxiety. When given in large doses, however, barbiturates knock people unconscious and repress breathing and other life-sustaining functions. They are also seriously habit forming. As a result, they are used today primarily in controlled circumstances—as general anesthetics in surgery, for example. Yet in the 1950s and '60s, doctors commonly gave their patients barbiturates to treat anxiety, emotional distress, and sleep problems. As the number of prescriptions increased, so did accidental and suicidal overdoses. Both Marilyn Monroe and Judy Garland died of overdoses of barbiturates. Unfortunately, doctors who wanted to dull their patients' emotional pain had few safe means for doing so.

The chemist Leo Sternbach changed all that. In the 1950s, he led a research team for the pharmaceutical company Hoffmann–La Roche in search of a less deadly tranquilizer. After years with no success, the company told them to cease their efforts. In a show of rebellion, Sternbach refused to clean up his lab, and it sat untouched for two years. A coworker sent to clear it out noticed a "nicely crystalline" compound amid Sternbach's mess. It turned out to be chlordiazepoxide, which testing revealed to have strong sedative effects without suppressing breathing. In 1960, Hoffmann–La Roche began to market it as Librium and then refined it over the next few years to create Valium (diazepam), in 1963, named after the Latin *valere* (to be strong).

Both drugs were tremendously successful, largely replacing older tranquilizers and sedatives by 1970. Medical professionals were thrilled; benzodiazepines were less dangerous and addictive than barbiturates. They could dull their patients' suffering without the risks and side effects. By the mid- to late 1970s, benzodiazepines were number one on "most frequently prescribed" lists, with 40 billion doses consumed annually around the globe. Valium was so popular that physicians referred to it as "V." Benzodiazepine prescribing peaked in 1978 and 1979, with Americans consuming 2.3 billion Valium tablets in each of those years. It entered the lexicon and drove an entire culture of chemical coping: the Rolling Stones immortalized "Mother's little helper" to "get through her busy day," while corporate jet-setters called it "executive Excedrin" because it eased the stress of frequent travel across time zones. Other pharmaceutical companies followed suit, developing and patenting their own benzodiazepines. The number has grown steadily, so that today almost thirty-five different versions are approved for use inside and outside the United States.

Companies flooded the market with benzodiazepines for fifteen years before researchers even understood how they worked (it turned out that they modify the main inhibitory neurotransmitter in the brain, gamma-aminobutyric acid, or GABA). With increased knowledge, clinicians' enthusiasm was replaced with caution as they witnessed more examples of the dependence, overdose, and abuse potential of the drugs in the 1980s and '90s. Hoffmann–La Roche noticed, for example, that its very effective sleeping pill, Rohypnol, another Sternbach molecule, had become better known as "roofies," the date rape drug. The company had to change the formulation so it wouldn't dissolve easily and would turn liquids blue, as a warning to potential victims.

Health care professionals were waking up to the fact that although drugs such as Valium, Ativan, and Xanax may have been safer than barbiturates, they were far from benign. Their danger was due to several factors. For one, benzodiazepines are nervous system depressants. And although they don't stop breathing and induce unconsciousness as readily as barbiturates do, they do significantly slow these functions, while also dampening higher-level decision making and motor control. Moreover, as their use increases, emotional dependence and physical addiction set in. People might find themselves taking more and more of the drug to get the same effect and start falling asleep at the wheel, slurring their words, losing their memory, and becoming confused. Even worse, when combined with other drugs such as opioids or alcohol, dangerous synergistic effects can lead to cardiac emergencies, coma, and death. A second danger is their potential for psychological addiction. Under their influence, people feel calm. Their emotional pain is relieved. There are few experiences as intrinsically rewarding as that, so the pull to take more of the drug—and find even greater emotional relief—is strong.

Once believed to be miracle drugs that saved lives and relieved suffering, benzodiazepines have since ceased to be the good guys of modern psychoactive pharmaceuticals. Despite that, they haven't gone away.

Benzodiazepine overdose deaths quadrupled between 2002 and 2015, a trend fueled by a 67 percent rise in prescriptions. Drugs such as Xanax are today a multibillion-dollar industry, in 2020 reaching $3.8 billion in sales in the United States alone. Using benzodiazepines briefly to manage an anxiety disorder in combination with therapy is a gold-standard treatment approach. But that's not what happens much of the time; more than 30 percent of American adults over the age of sixty-five take benzodiazepines longer than prescribed, and around 20 percent of younger adults do the same. Because the calming effects of benzodiazepines can be felt in a single dose—unlike those of other medications, such as antidepressants, which require continuous use for a month or more to feel the benefits—taking a benzodiazepine pill to "take the edge off" has become a lifestyle. The longer benzodiazepines are used, the more likely we are to become emotionally and physically dependent on them—and the harder it is to get off them. Physical withdrawal and a rebound of anxiety and nervousness are common when people are weaned off the medications. It drives many of them back to taking the pills.

Despite the growing awareness that these drugs are habit forming and potentially hazardous, it's easy to ignore the red flags and warning signs of addiction. We don't see ourselves as addicts when we have a prescription or when we take a drug just when we "need it."

To understand the potential danger of benzodiazepines, we should consider the proliferation of another type of painkiller: opioids. Benzodiazepines and opioids are often taken together—the former for emotional distress and the latter for everything else. It's not that doctors prescribe them together; they actively warn patients *not* to take them at the same time because of their dangerous synergistic effects, which increase the risk of overdose deaths. The National Institute of Drug Abuse reported in 2019 that the number three prescribed class of drugs associated with overdose deaths was benzodiazepines; numbers one and two were the opioids oxycodone and hydrocodone.

How did we get to this point, where pain relief is the number one cause of drug overdose deaths?

The Business of Dulling Pain

If we need any further evidence of our societywide desire to eradicate all pain—physical, emotional, and psychological—we need look no further than the opioid crisis. In the pursuit of dulling discomfort, millions of us ended up suffering more than we could ever have imagined.

Opioids work by attaching to receptors in our cells and releasing signals that effectively muffle our perception of pain and boost our feelings of pleasure. They have been regulated by the Food and Drug Administration (FDA) since the early twentieth century to treat acute pain and cancer pain. But they went from being a target of caution, given their readily acknowledged abuse and addiction potential, to causing a deadly epidemic in the twenty-first century.

At the height of the prescription painkiller crisis, there were enough pills available in the United States for half of the US population to have one—twice the volume of opioids considered normal by public health officials before the prescription boom began in the late 1990s. To put that into perspective, the United States, which has 5 percent of the world's population, consumed 80 percent of the world's prescription opioids. From 1999 to 2019, nearly 247,000 people died in the United States from overdose due to prescription opioids. In 2019 alone, more than 14,000 people died, an average of 38 each day—more than half of whom were teens.

That was on a scale never seen before. Deaths involving prescription opioids alone more than quadrupled from 1999 to 2019. The victims didn't seem to fit our collective image of the "type of person" who died from an overdose. Opioids were killing our mothers, fathers, brothers, sisters, and children. They were killing the most famous among us: Heath Ledger in 2008, Michael Jackson in 2009, and Prince in 2016. In 2017, the US Department of Health and Human Services declared that the abuse of opioids, lumping prescription painkillers and heroin together, was a public health emergency.

What had changed? One simple thing: the pharmaceutical industry. Purdue Pharma, the manufacturer of the most ubiquitous of the prescription opioids, OxyContin, almost single-handedly engineered the opioid crisis. Not only did the company bribe doctors to prescribe the drug, wooing them with free trips and paid speaking engagements, it also made false claims that its "slow-release" formulation had low abuse potential—despite the fact that scientific evidence backed the opposite conclusion. Doctors closed their eyes and kept writing prescriptions. Purdue Pharma was fully aware that OxyContin was being frequently abused, including "reports that the pills were being crushed and snorted; stolen from pharmacies; and that some doctors were being charged with selling prescriptions," according to Barry Meier of the *New York Times*. Yet the company continued and even accelerated those practices. Legal suits stopped Purdue Pharma and the Sackler family, who owned and controlled the firm, from continuing their predatory practices. But billions of dollars in fines couldn't erase the damage they had done.

As with the proliferation and danger of benzodiazepines, the opioid crisis is a direct reflection of how persistently medications—for emotional and physical pain—are pushed on us and how amenable we are to accept the solutions they appear to offer. The opioid crisis was in many ways the apotheosis of our decades-long march toward rejecting all experiences of pain. When it came to the explosion of benzodiazepine addiction and deaths, however, there wasn't a big bad pharmaceutical company at the helm. Nothing so dramatic. But benzos actually started what opioids completed: the acceptance of chemical calm. Though physicians aim to relieve suffering, they forget—or never knew—that anxiety isn't the kind of discomfort that should be eradicated. It should—and must—be engaged and addressed if it is to be safely relieved and put to good use.

"I Felt like Superman"

Centuries of history convinced us that anxiety is a disease. Decades of our health care system persuaded us that when we're in emotional or physical pain, we should pop a pill. To understand what this means for our future, we have to turn to the bellwethers of the years to come: teenagers.

In any given year, 18 percent of adolescents will struggle

with debilitating anxiety. In the United States, that currently equals around 40 million kids. They are well aware of their struggles; a Pew Research Center Report released in February 2019 showed that 96 percent of teens surveyed believed that anxiety and depression were a significant problem among their peers, with 70 percent saying that it was a major issue. The tens of millions of people diagnosed with an anxiety disorder before their eighteenth birthdays are also much more likely to suffer in adulthood from continued anxiety, depression, addiction, and medical problems. Teen anxiety is the gateway to the present and future health—both good and ill—of our society.

Clearly, something has changed, and many of us, whether we're parents of teenagers or not, believe that we can't afford to ignore the signs anymore. At the same time, we compound the problem by telling discouraging stories about these kids: that members of Gen Z, along with Millennials, are emotionally crippled, coddled, lazy, and screen addicted. But maligning them is just a way to bury our fear; we are afraid that our future citizens and leaders will be constitutionally incapable of coping with the world we will hand them. We are also afraid that kids' anxiety will block them from succeeding in the increasingly competitive world in which, many people argue, the American meritocracy-work-hard-and-reach-the-heights-of-success fever dream is in its death throes. A student at a gifted and talented high school in Manhattan put it this way: "Adults make us go to the counseling office as soon as our grades slip or if we start to get nervous about taking tests. I think it just makes them anxious when we're anxious. They're scared we're going to mess up."

Kids have heard the message: get your anxiety under wraps, double time. What better option than chemical control?

Indeed, the cliché of the stressed-out suburban house-wife secretively downing her Valium (and maybe a martini) to get through the day has been replaced by the stressed-out teen gulping Ativan and Xanax from his or her school locker. Feeling anxious about that test? Just pop a Xani. The rush to dull our feelings with pills has made the world a more dangerous place, especially for young people.

You can see evidence of that trend in an unlikely place. In 2019, the website Complex blew the assumptions about the victims of the benzodiazepine crisis out of the water with an investigative piece called "Bars: The Addictive Relationship with Xanax & Hip Hop." In the video were stories of musicians and their friends who had become dependent on Xanax and other benzodiazepines in an effort to medicate their anxieties out of existence. As one man said, "I felt like Superman. I normally feel anxious, but when you're on it, you feel like no one can stop you." So common had the drugs become that in the mid-2010s a rapper took his stage name from his drug of choice: "Lil Xan," as in Xanax.

Eighteen-year-old Jarad Anthony Higgins, known pro-fessionally as Juice WRLD, was no gangsta rapper; he was vulnerable and brutally honest about his emotions. In his song "Righteous" Juice WRLD shifts within seconds from describing how powerful he feels in his white Gucci suit to how he self-medicates with "five or six pills in my right hand, yeah/Codeine runneth over on my nightstand" to deal with "My anxiety the size of a planet." The solution doesn't work because, as he explained in another song about medicating away emotional pain, "Bad Energy," "Can't explain this feeling/Kinda feels like I'm losing/Even though I'm winning."

The terrible tragedy is that he did lose. By the end of 2019, Juice WRLD and several other high-profile emo rappers, including Lil Peep, had died from overdoses of benzodiazepines and painkillers—at the age of twenty-one.

As the lights dimmed and the curtain rose in the Lyceum Theatre on Broadway, the actor Will Roland, playing the awkward and painfully anxious teen Jeremy Heere, sang the opening lyrics of "More than Survive"—along with what sounded like half the audience, who knew every word of the song about anxiously staring down the barrel of another miserable day of high school: "If I'm not feeling weird or super strange/My life would be in utter disarray/'Cause freaking out is my okay." It was 2019, and that was the opening song of the musical *Be More Chill*.

The plot centers on Jeremy, an anxious, twitchy, social misfit nerd, who is offered a computerized "pill," called a Squip, which rewires his brain to "be more chill" and fit in with the popular kids. You don't have to think hard to see how taking a Squip is the digital version of popping a Xani.

The Squip "helps" Jeremy with his anxiety by dictating exactly what he should do to win friends and influence people. It does so in a form visible only to Jeremy—that of

Jeremy's ideal cool guy, Keanu Reeves's character in *The Matrix*. Drama soon ensues. Everyone who takes a Squip to be more chill—their numbers are exponentially growing—eventually becomes a zombie/*Invasion of the Body Snatchers* pod person and even worse—if that's possible—will soon "glitch" and go into a coma. Jeremy learns that people will do anything, even risk their lives, to erase their anxiety.

How did this idiosyncratic show create so much fan love? I believe it held up a uniquely honest mirror to young people's lives and gave them a choice about a way forward: you'll be handed anxiety-erasing Squips, but you don't have to take them. You can feel weird, you can even freak out, and still be okay.

Why can't the rest of us send this same message to kids? Because we've bought into the story that erasing anxiety and becoming comfortably numb is the best—and perhaps only—solution. And it's not just through medications. We and our kids have become consumed by one of the most powerful tools ever created to avoid and escape anxious, uncomfortable feelings. It's right at our fingertips, in the palm of our hand.

6

Blame the Machines?

Anxiety and digital technology seem inextricably linked. Although we often assume that too much screen time and social media *cause* anxiety, the connections between these ubiquitous aspects of modern life are more complicated than that.

On the one hand, devices allow us to escape our anxieties and worries; within seconds, we can take refuge in a universe of options: playing a distracting game, reaching out to Dad, shopping for a new garden hose, streaming a favorite TV show, or getting some work done. On the other hand, when we become absorbed in our screens, research shows, we often end up feeling more anxious, isolated, and exhausted than we did before we started. That's especially the case when we feel compelled to obey the rings, dings, and notifications telling us to check our social media feeds; when we reach for the phone next to our bed the moment we wake, as a smoker would for a cigarette; when even in the briefest moments of quiet, boredom, or distress, we feel the urge to swipe through an infinite scroll of information.

This is why we've come to think of phones as addictive. Unlike drugs, however, devices don't necessarily trigger the hallmarks of addiction, such as tolerance, when we need to use more of something to get the same effect, or withdrawal, painful physical symptoms that arise when we stop using them. Yet leaving aside whether the addiction metaphor is accurate or not, digital technologies aren't so different from benzodiazepines; we use them to escape the pain of the present moment, but if we get hooked on them, we end up feeling worse. And like the drugs that produce a chemical calm, devices can keep us from seeking out beneficial ways to manage our anxieties. Here's how they manage that: first, by giving us an alluring escape from anxiety—at least temporarily; and then, by careful design, encouraging us to come back for more, even when the escape no longer works.

The Ultimate Escape Machines

When we're anxious, we gravitate toward experiences that dull our unpleasant feelings. What serves this goal more immediately, more easily than mobile devices? We use these little escape machines, nestled in pockets and purses, clutched in our hands wherever we go, in countless ways. They take us out of the experiences we're having in the present and convey us somewhere else. That can't be all bad. Yet when we habitually evade anxious feelings, the avoidance paradox kicks in, and our anxiety is likely to increase.

But not all digital time is created equal, and whether

digital technologies ramp up our anxiety or not depends on *how* we use them.

Take social media, one of the best-researched aspects of our digital lives. There are two ways we can use them: actively and passively. Active use is the purposeful sharing of "content"—anything from texting with a friend or arguing with one's arch nemesis over Twitter to sharing photos with family members or posting the latest video of you killing it on the ukulele for your sixty-three followers to watch. Passive use, on the other hand, lacks all this creativity and panache. With passive use, we don't have to share our personality or talents, don't have to express thoughts or feelings, don't have to commit to a belief. We just casually consume—browsing the web, scrolling through social media feeds, or reposting content from others. That seems harmless enough, at worst just a time waster. Or perhaps it's like munching on potato chips—mindless, effortless, but before you know it, you've eaten the whole bag and have only a stomachache to show for it.

But does the way we use social media make a difference? A decade of research has turned up some answers. They're not so simple.

A large-scale survey of more than ten thousand Icelandic teens showed something possibly significant. Researchers asked them to report all the active and passive ways they used social media in the course of a week, as well as symptoms of anxiety disorders and depression. It turned out that when they spent more time using social media passively, they were also more anxious and depressed—even when they felt socially supported by others and had strong self-esteem. Conversely, when they spent more time using social media actively, they were *less* anxious and depressed. It didn't matter how long they spent bent over their social media feeds; what mattered was what they were doing there.

Despite the fact that this study involved an impressive number of people—and that its basic finding has been replicated at least a dozen times—it remains correlational. In other words, we still don't know if social media use *causes* anxiety or depression. It could easily be the other way around: people who are more anxious or depressed may be more likely to consume social media passively because doing so is undemanding or relaxing. Or some other factors that the researchers didn't measure—trauma, family circumstances, genetics—could be causing the increase in distress. Are we any closer to knowing which way the causal arrows point?

In 2010, researchers from the University of Missouri and Columbia University wanted to take a first step toward answering this question. College students came into the lab and were asked to do something familiar: hop onto Facebook and use it as they normally would. Only later were they told that the researchers had been tabulating their every point and click—specifically how much time they had spent passively browsing around versus actively seeking out information and communicating with friends. At the same time, the researchers tracked the participants' positive and negative feelings. But instead of asking the participants how they felt, the researchers used a bias-proof method: facial electromyography, or the strength of the electrical activity in the muscles involved in smiling (orbicularis oculi) or frowning (corrugator supercilii).

Neither passive nor active use increased frowning, which presumably indicates negative feelings. But passive use directly reduced smiling, which suggests that it doesn't make us any happier. Of course, less smiling doesn't automatically equal more anxiety or depression, but at this stage of the science, this study is one of the few showing that different uses of social media actually cause any reaction at all. This gives us a sense of how much we *don't* know.

But let's assume for the moment that this research is correct. If using technology in passive ways actually dampens positive feelings, why do we keep going back for more?

The One-Armed Bandit

Sometimes digital technologies seem so perfect, so effortless, that we assume that their design is inevitable. But cleverness of design can fool us into forgetting that nothing about the way we consume these technologies is inevitable.

Devices, websites, and social media platforms are purposefully and relentlessly engineered to keep us looking at their screens and to hook us into opening up yet another app. How? They are designed like casinos loaded with slot machines.

The infinite scroll is a perfect example. As we swipe down

the screen, information continuously pops up as we go, so that we never need to stop, click, or wait for the next page of information to load. This removal of pauses gives us fewer opportunities to stop and consider "Is this what I want to be doing right now?" We go on automatic pilot, doing what feels good in the moment. Indeed, research shows that the simple act of repeatedly swiping and scrolling temporarily calms and soothes us, makes us feel good, and might even temporarily reduce biological stress, measured through skin conductance or subtle changes in blood flow beneath the surface of the skin.

The design of casinos plays on the same principles of automaticity. A casino walkway, for example, has no right angles, only gentle, winding curves, so that it's easier to wander from one game to another, letting our impulse to play and win move us along. No pauses are required. And like a casino walkway, the infinite scroll encourages us to keep moving, blithely swiping until we come to the intended goal: the games of chance.

Our devices, and many of the things we do on them, are designed like little slot machines. That's because, like a slot machine and all other types of gambling, they provide intermittent, unpredictable rewards. These kinds of rewards do an outstanding job of encouraging and reinforcing whatever behavior leads to them. People get hooked on slot machines because they never know when they're going to hit the jackpot, the three-cherries-in-a-row payout, so they keep pulling away. Likewise, to keep people picking up their device, clicking, scrolling, buying, and posting, unpredictably and intermittently reward them with "likes," news, drama, or excitement.

Smartphones keep us coming back for more because we never know when we'll get three cherries in a row, whether it's a message from a friend, news we've been waiting for, or a funny cat meme.

Conversely, the concept of doomscrolling is a "perfect" marriage of the casino floor of the infinite scroll and slot machine reinforcement. Doomscrolling is something we've probably all done when we're anxious—scrolling obsessively through bad news, even when it distresses us. Though doomscrolling certainly existed—in practice if not in name—before the covid-19 pandemic hit, the use of the phrase skyrocketed during quarantine—that's when Merriam-Webster online added it to the "words we're watching" list. It's not hard to picture all the hours we spent glued to our screens as we consumed every bit of news we could about the virus, partisan politics, racial injustice, unemployment rates—anything negative or distressing, we can doomscroll it.

But along the way, while doomscrolling, we might get a few rewards—a nice text from a friend, a happy piece of news amid the wreckage of current events. That's just enough to keep us going in the pursuit of feeling better.

Yet doomscrolling is actually an attempt to manage our anxiety; by gathering more and more information, even when it's bad, we hope to reduce our uncertainty. This is a good strategy under normal circumstances, but unfortunately the digital world is not all that "normal"; it prioritizes negative information over positive, polarizes us into information bubbles, and rewards the sensational over the factual, the Twitter trolls over kindness and calm.

Doomscrolling is far from the only mindless way we use technology in search of relief from anxiety. What do repeatedly clicking on a cartoon image of a cookie and guiding a colored ball through like-colored obstacles have in common? These are wildly popular hypercasual games: Cookie Clicker and Color Switch. Hypercasual games are by definition fun, simple, repetitive, and absorbing. Some are challenging, but many use simple game mechanics that require so little attention that they are played mostly while people are doing other things such as watching TV or eating. Talk to hypercasual game users, and they'll tell you that they play to relieve stress and anxiety, unwind after a long day, and distract themselves from worries. Many people use them to go to sleep.

A few scientists have studied hypercasuals as interventions for anxiety, positing that these little games soothe people by triggering a sense of flow with each fluid, relaxed, repetitive action. Like the infinite scroll, they seem to lull us into a calmer, automatic-pilot state. The scientific jury is still out on whether that's helpful in the long run; if we scroll to avoid confronting distressing emotions, it probably isn't. But the idea that they might be helpful tracks with earlier research showing that simply scrolling through social media feeds temporarily dampens biological stress. In 2021, these little games were big business with millions of people playing them, often for hours at a time.

Certainly we've long used entertainment technologies to relax—television and radio leap to mind—because they suck us in, absorb our attention, distract us from worries. TV wasn't called the "boob tube" for nothing. But what's new is that the most powerful tech companies on the planet now want us to pay attention to our devices *all the time*, so that they can collect massive amounts of the most valuable digital product in the world: our "personal data," meaning what we believe, what we want, where we go, and what we do. That's why digital technologies are designed like casinos, hard to get out of. It's smart business.

These radical and unprecedented efforts at commodifying our attention are relevant to anxiety because they work only when our eyeballs are locked on our screens. And when our eyes and minds are stuck there, we may lose opportunities to benefit from one of the best tools we have for managing anxiety: our real-life social connections.

Social Brains in a World of Screens

Maneesh Juneja is a digital health futurist; he imagines how emerging technologies could make the world a happier, healthier place. Sounds like a great job if you can get it. But after a terrible loss—his beloved sister passed away unexpectedly in 2012—he awoke to a surprising reality: it was only face-to-face human connection that helped him cope with his grief. Indeed, although his life revolved around digital technology, it was the last thing he turned to for coping with his loss. A virtual reality backyard barbecue left him feeling disconnected and worse than before, whereas simply going to a human cashier at the local grocery store rather than the much faster self-checkout line buoyed his mood. Long before Zoom infiltrated our lives, Juneja realized that although technological means of connecting are immensely valuable, something about human presence—touch, eye contact, and voice—is uniquely healing.

This makes it all the more ironic that social media—perhaps one of the greatest misnomers of our time—often block us from leveraging human presence to relieve our anxiety and distress. We already know that having strong social support strengthens our health and that loneliness and isolation can take years off our life expectancy. How does this work? One way is that during times of stress, the presence of a supportive loved one changes our biology. The hand-holding neuroimaging study we saw in chapter 2 showed that the presence of loved ones literally gives us more brainpower to cope with threats. Can this benefit be conveyed through technology when we can't hold hands? In 2012, researchers from the University of Wisconsin–Madison wondered about that, too.

When we benefit from social support in person, our level of the stress hormone cortisol plummets, while production of the social bonding hormone oxytocin increases. But do these same powerful biological effects emerge when social support is provided through technology? Let's look at the relationship between mothers and their teenage daughters. In one study, girls first endured the anxiety-provoking

Trier Social Stress Test. After giving a nerve-wracking public speech and performing a difficult math problem in front of judges, the teens felt understandably rattled. They were allowed to reach out to their moms in one of three ways: in person, on the phone, or via text. A final group of teens sat alone and received no support.

For their part, the moms were told just to be as emotionally supportive as possible. When they provided that support in person or on the phone, their daughters' stress hormone levels dropped and their social bonding hormones rose, as expected—signs that the support was working. But when the teens received comfort from their mothers via text, nothing changed—the girls showed little to no release of oxytocin, and their cortisol levels were as high as of those who received no support. Connection through digital devices just wasn't the same as the comforting voice or physical presence of a mom. This suggests an evolutionary mismatch: that perhaps we humans benefit most from social support when we perceive unmediated human presence.

A second way that social connection works its magic on anxiety is through another sensory experience: eye contact. Unlike almost every other animal, even our closest primate cousins, only humans have the ability to share meaning and intentions by locking gaze. In other words, we communicate simply by making eye contact with each other. We also find solace. Imagine two people sitting quietly together. They turn, look into each other's eyes, and wordlessly understand one another. From the earliest days of life, children can do the same. Babies look into the eyes of their caregivers to seek comfort, learn the back-and-forth reciprocity of play, and observe how their own feelings and actions affect others. As we grow, we build on these skills to eventually become expert in the subtle nuances of social communication.

You can see the importance of human gaze in how the human eye evolved. The whites are much larger than in primates and other animals. This allows us to track and coordinate the direction of other people's gaze with exquisite accuracy; it's easier to see which way our pupils point when the irises are surrounded by white. When we can follow one another's gaze, we also can better understand what the other is doing, wants, and wishes for us to do. Some scientists argue that this seemingly simple characteristic was fundamental to *Homo sapiens'* evolutionary advancement as a species because it allows us to effectively cooperate and coordinate our goals and intentions.

If we chronically disappear into our screens, our head and eyes down, do we risk weakening this crucial channel of human communication?

In 2017, we explored this question in the context of a key relationship—between parents and their young children. The subjects started the research session by playing together as they would at home. After they had settled into a nice groove, the parents were instructed to unceremoniously interrupt the play by bringing out their mobile devices. To make sure they ignored their children and kept their eyes on their screens, we asked them to fill out a short questionnaire on-screen. After a couple minutes of that, the parents were directed to return their attention to their children and resume playing.

Snubbing someone in favor of a mobile phone may be common in many families; it even has a name, *phubbing*. But unsurprisingly, the children in that study still showed distress and made strong bids for attention when their parents were busy on their phones. Their negative emotions tended to linger into the reunion period, and although many of the children bounced back and happily reengaged with their parents, others remained anxious and preoccupied. They seemed worried that their folks might disappear into their phones again.

The kids who were used to phubbing didn't fare any better. Indeed, the parents who reported using screens in front of family members more often had children who were less able to bounce back emotionally during the reunion. Those kids showed less positive emotion and more negative emotion and took longer to resume playing, even when their parents returned their full attention to them.

We recreated this study in 2019 for a network TV special report, *ScreenTime: Diane Sawyer Reporting*, and had the chance to delve more deeply into how children perceived the loss of their parent's gaze. One boy reacted immediately, repeating seven times, with increasing loudness, "We have some other things to do, Mommy. Mommy, stop, Mommy, it's time to go." A girl who had been happily playing with her mom just moments before, silently pulled up a chair and sat facing her parent when the screen came out. Instead of busying herself with playing or trying to get her mom to play again, the little girl just waited, stock still, unsure when her parent was going to come back to her.

The message of the study wasn't that using devices in front of our children and family will damage them. But our findings suggest that if we constantly disappear when we're with our loved ones, we may lose opportunities to connect in the ways that benefit us all.

In a second study, we tested the impact of phubbing on adults. We assigned pairs of subjects to work together on a difficult puzzle. One of the adults in the pair—a research assistant who was posing as a participant—continually interrupted the task by breaking eye contact, texting, and talking on the phone. In the control group, the pair worked together to solve the puzzle without interruption.

Like the study with parents and children, the effects of breaking reciprocity and connection through eye contact were far from trivial. Adults not only found being phubbed by their problem-solving partners to be rude, they also showed more anxiety.

What if the Kids Are All Right?

If we are to believe the headlines about digital technology, we must choose between one of two camps: the doomsayers, who tell us that smartphones shorten our life spans and cause everything from teen anxiety to suicide; or the naysayers, who tell us that all the panic is unnecessary and our hysteria about digital technology will fade much as it did for past generations' worries about watching too much television.

Is there a middle way?

To figure that out, we need to talk to the people who are really in the know: digital natives. In an NPR report from 2018, "Teen Girls and their Moms Get Candid About Phones and Social Media," it was clear that kids feel torn—torn between knowing that social media can sometimes make them feel anxious and depressed and knowing that their phones provide social connection and emotional relief they feel they can't live without.

"Adults don't know how important the phone is for teenagers," one adolescent said. "I feel like, when you do have social media and a phone it makes you more friendly. I sit next to some boy in one of my classes. He doesn't have a phone. He won't talk the whole class. It makes you anti-social."

"I don't necessarily enjoy being on it," said another. "—Well that's so not true. I enjoy being on it, at the same time, I also know what it's doing to me. I know that it actually causes me a lot of anxiety. But again like, it's really easy. I can sit on a couch, not move my body, hold something in my hand, and do so much. I can exist in another world without doing anything."

We can all relate, especially after the pandemic, during which screens became not only our lifelines but also our Zoom-fatigue, infinite-scrolling banes of existence. At times, we felt addicted to screens and to social media in particular.

But the addiction analogy is oversimplistic. The reward centers of the brain may be active when we feel hooked on Instagram as well as when we're physically addicted to benzodiazipines, but they're also active when my excessive love of salt and vinegar chips is triggered and it's hard to say no to the next chip. Moreover, many of us get sucked into using social media for reasons that have much less to do with reward: complex social motivations, information gathering, and professional goals, to name a few.

Some researchers continue to ignore the nuances of our relationship with digital technologies. They've decided, in the absence of proof, to trumpet the attention-grabbing headlines that smartphones are addictive, have psychologically destroyed a generation, and are fueling the epidemic of teen anxiety and suicide in the United States.

Yet the fact remains that there is almost a complete lack of direct evidence that devices actually *cause* significant mental health problems or that using social media *makes* us anxious. One study, based on survey data from hundreds of thousands of teens, concluded that a spike in anxiety and depression among youths in 2011 was likely due to broad smartphone adoption around the same time. Using the same data, however, a separate research group from Oxford University showed that eating more potatoes than average was just as strongly associated with an increase in anxiety—reminding us that correlation never equals causation.

In one of the few prospective longitudinal studies of social media use and emotional adjustment—that is, researchers first measured social media use and then tracked whether it predicted well-being over time—Sarah Coyne at Brigham Young University and her colleagues found no associations between the time spent using social media and anxiety and depression over the course of eight years, spanning early adolescence into young adulthood.

Even these findings are far from conclusive. We won't know anything for certain until we focus our research efforts on the hard questions: What types of social media use help and what types harm? Can our biology help us understand why we are affected, if in fact we are? Which of us are resilient and which are vulnerable? And does the impact of digital technology change over time as we ourselves change?

Almost ten years after Leslie Seltzer and her colleagues brought teen girls and their moms into the lab to study social support, we invited teenagers and their best friends into our lab and divided them into three groups. Two of the groups were simply asked to discuss things that were bothering them and then give one another emotional support— one group via Zoom, a second by text. The third group sat alone to think about what was bothering them. After the conversations, rather than measure stress hormones and social bonding hormones, we measured teens' brains using EEGs while they viewed emotionally intense pictures such as of a very sick person in a hospital or a soldier caught in a violent altercation. Our theory was that the teens who felt the most socially supported would be better able to manage their emotional reactions to the photos. And we figured that Zoom would be the most effective way to support a friend:

seeing each other's faces, hearing their voices, tuning in to how the person feels in real time.

But that wasn't what we found at all. The group members who texted one another showed the calmest brains. Even more interesting, the brains of the teens on Zoom looked identical to those of the teens who were left alone and unsupported.

We were stumped—not just because that wasn't what we had expected but because it seemed to contradict the 2010 study of moms and their teen daughters who showed no benefit from texted support. So we talked to the teens. This was 2019, so most of them had grown up texting, and they preferred it to other forms of communication. The slang of textese, emojis, and gifs was safe from the prying eyes of adults, who couldn't understand half of it, but for the teens it was a rich and complete vocabulary. Zoom left them cold; it seemed out of sync, awkward, not really like talking in person. They weren't against face-to-face time; they still craved being together with their friends, in person. But they also liked being able to pause during a text conversation to think through what they want to say, to absorb the distress a friend feels and that they might feel themselves. On video chat and face-to-face, for that matter, people have to react instantaneously; they don't have time to consider. From that perspective, texting helped them be the best, most supportive friends they could be.

While growing up, I never had to worry about internet trolls and haters and algorithms. I didn't even know what an algorithm was. As a young, developing person, I didn't feel the constant heat of the social media spotlight as I moved through my self-conscious teenage years. I'm not so sure I would have coped particularly well.

But Gen Xers like me should hold our assumptions in check. I learned this personally after receiving an email from a student after the student read a *New York Times* op-ed I wrote, in which I called for more nuanced discussions about social media and mental health rather than assuming that digital technology is a simple and direct *cause* of problems such as teen anxiety.

Dr. Dennis-Tiwary,

I am currently enrolled in a Concurrent Enrollment English 1010 class. I am writing to inform you that your article "Taking Away the Phones Won't Solve Our Teenagers' Problems" is extremely appreciated among the youth of this course. We have been assigned to read, annotate, and summarize this article. For the last few months, we have read nothing except for anti-technological articles. This article of yours was a huge relief for us. We felt as though someone actually understood the digital natives of this world.

From all of my classmates and I,
Thank You

PART III

How to Rescue Anxiety

7

Uncertainty

Uncertainty is the only certainty there is, and knowing
how to live with insecurity is the only security.

—John Allen Paulos, *A Mathematician
Plays the Stock Market*

It's the human condition. Every day is a set of probabili-
ties, a gamble that what usually happens is likely to happen
again—we'll wake up in the morning, do the things we
planned to do that day, head back to home base to eventually sleep, only to wake up the next morning to start the
gamble all over again. But of course nothing about life is a
sure bet. Most of us intellectually and abstractly accept this
to be true, but few of us dwell on it. When we do come face-
to-face with life's uncertainty, we feel a tension, a discord
between our assumptions and reality. Something untrust-
worthy and unreliable has entered our lives. It's this tension
that makes us sit up and take notice, because we know that
whatever happens next could be terrible or wonderful or just
meh—and so we need to do something about it if we can.

In other words, uncertainty is possibility. Even thinking about it propels us into the future.

It's the end of July 2021, and I wake up feeling stuffy and with a headache and sore throat. Probably just a summer cold or allergies—or could it be covid? When it doesn't go away after a day, I take an at-home test for the virus. As we're waiting for the results, my husband is pacing. He survived covid and fears any of us coming down with it—especially since our daughter is too young to be vaccinated.

I know I *could* be infected, but I think the chances are slim. My husband and I both feel uncertainty, but we are at different points on the spectrum: he's leaning toward the negative, while I skew positive. But both possibilities are still on the table, meaning we have some control over the future: I can take a test, track my symptoms, quarantine myself, and take precautions so that my daughter doesn't get sick. This is the sweet spot of uncertainty; it offers the chance to exert some influence over what happens next.

Say No to the Dark Side

A long time ago, in a galaxy far, far away, a boy named Anakin Skywalker was born on a desert planet. An ancient prophecy foretold that he would bring balance to the universe by uniting the light and dark sides of the Force. Instead, he was seduced by the dark side. This, of course, is the beginning of the *Star Wars* saga, the premier speculative fiction mythology of the past century. To some, it borders on a religion. To me, it's a parable of why we need uncertainty.

Anakin succumbed to the dark side because he became obsessed with preventing his greatest fear: the fact that one day, his beloved wife, Padmé, would die. But it wasn't the certainty of death that tortured him; it was the uncertainty of *her* death—he couldn't abide not knowing how and when she might die and whether he'd be powerless to save her, just as he had been unable to save his own mother from early death at the hands of raiders. And when Padmé did in fact die in childbirth—and Anakin was deceived into believing that he himself had killed her—what had started as his rejection of uncertainty became intolerable grief and rage. Anakin was soon transformed into the most iconic bad guy in modern cinema, Darth Vader.

Anakin's true downfall wasn't his love for Padmé. It wasn't even his fear. It was that he couldn't accept uncertainty. He saw only certain disaster, failing to see that he and Padmé could have had a long, fulfilling life together—a goal he could have worked toward. Because he lost the ability to imagine positive possibilities—that is, because he lost his uncertainty—the dark side consumed him.

The moral of the story? Rejecting uncertainty means rejecting the potential for tragedy as well as for joy. Also, don't be Darth Vader. Luckily, our brains have evolved to help prevent that from happening.

Our Brains Seek Out Uncertainty

Uncertainty is a key to survival. From an evolutionary perspective, it's not the certain threats that are most dangerous, it's the unknowable ones. This limits our ability to prepare for them, learn from them, and actually do something to— you know—survive.

Accordingly, our brains don't ignore uncertainty; they lean into it, so much so that evolution has designed the human brain to automatically and effortlessly pay attention to the unexpected, the unpredictable, and the novel. It's called the *orienting response*. It's reflexive and unconscious, so we can't stop ourselves from doing it even if we try. It's like our lower leg popping up when the doctor hits our knee with a little rubber hammer—but lightning fast. Our brains have evolved to be uncertainty radars.

Indeed, you can observe the orienting response in the form of brain waves. Imagine a computer task where you have to push the "up" button on the keyboard if you see a Y on the screen and the "down" button if you see an N. The Y's and N's come fast and furious, so sometimes you get it right, and sometimes you get it wrong. The computer chimes pleasantly when you succeed, but when you fail, *braaap!*, an annoying buzzer goes off. And once in a while, you get a neutral *ding-ding*. Did you get it right or wrong? You're not sure.

Dozens of studies using tasks like this have shown that within just a third of a second, our brains respond to feedback with specific changes in electrical activity—otherwise known as brain waves—that can be measured with an electrocardiograph (EEG). We've given these waves jargony names such as "error-related negativity," "error positivity," and "feedback-related negativity." They simply signify that our brains are calculating: Did I get it right, did I get it wrong, or is it uncertain?

When the brain waves grow bigger, that means more energy and power are being expended by our neurons. And what causes some of the biggest brain waves? Uncertainty, that ambiguous little *ding-ding*, especially when we're feeling self-conscious or stressed. Don't get me wrong, mistakes elicit large brain responses, too, especially in comparison to when we are correct. This makes evolutionary sense, because survival often depends on being able to learn from our mistakes rather than just luxuriating in being right. But our brains track uncertainty with extra vigor because that's what we really have to figure out.

That requires brainpower, or what psychologists call *cognitive control*, the ability to learn, decide, and change our thinking and actions to solve problems. Luckily, at the same time that our marvelous brains are focusing on uncertainty, they are also revving up our powers of cognition. Indeed, few things cause as much stress to a human brain as being out of control. Just take a 2004 meta-analysis combining data from more than two hundred studies. The studies examined the types of situations that cause the most stress— everything from being negatively judged on public speaking to completing difficult mental tasks such as arithmetic to being inundated with continuous loud noise.

When the studies were considered in relation to one another, what was the situation that triggered the highest stress hormone response? None of them. It didn't matter what the situation was; what counted most in all the studies was the degree to which the situation was uncontrollable by the participants, especially if it had to do with other people; having to perform in front of a judge who gave you the thumbs-down no matter how well you performed was among the most stressful, for example.

How do our brains rev up cognitive control when faced with uncertainty? They do it by prioritizing the perceived uncertainty over pretty much everything else. In one study, for example, participants completed a tricky perceptual task: they studied two images and had to decide whether one was more pixelated. Though some were obvious, others were more difficult; images that were subtly different were hard to tell apart. For any of the pairs, the participants could refuse to answer and instead indicate that they were uncertain.

Brain scans showed that when participants said they felt uncertain, a broad network of neural regions underlying cognitive control was activated, such as the prefrontal cortex and anterior cingulate cortex. In contrast, when the participants had to make a difficult decision between two similar images, brain scans showed that the cognitive control regions were only weakly activated. In other words, whereas uncertainty caused the cognitive control cavalry to come thundering in, solving a tricky problem barely got the riders onto their horses.

This is the miracle of uncertainty—without our expending any conscious effort, our brain does two things surpassingly well: it notices the uncertainty, and then it does everything it can to control it. This is what enabled humans to learn, adapt, survive, and thrive across tens of volatile and unpredictable millennia.

We learned this lesson the hard way recently, when we ourselves became unwilling participants in a collective case study of uncertainty.

The Pandemic of Uncertainty

Because you lived through covid-19, you experienced uncertainty in its rawest form: Will I die? Will my loved ones? Is it safe to leave our home? Will I still have a job in a few months? Will the overburdened health care system be able to take care of us if we get sick? Will the global economy collapse? How long will we have to endure social isolation, distance learning, Zoom fatigue?

We experienced a pandemic of uncertainty, and it was 100 percent contagious.

The pervasive unpredictability of the pandemic at times felt like torture. Psychologists call this *intolerance of uncertainty* and measure it by asking people if they agree with statements such as "Uncertainty keeps me from living a full life," "I always want to know what the future has in store for me," and "I can't stand being taken by surprise."

Despite these understandable feelings, evolution has

prepared us for the world falling apart, so we didn't just sit on our hands and wait for the virus to get us. Uncertainty inspired us to take action. We did lots of things.

Take mask wearing. At first we were told to save the available masks for the frontline health workers. So when we couldn't find any for love or money, we sewed them out of old T-shirts or used handkerchiefs. Once masks finally became available, we wore them religiously and coveted them like treasures. I learned that a pal of mine was a true friend forever when he offered me one of his N95 masks.

We realized that even wearing a mask was no guarantee of safety, but uncertainty kept us believing in doing so anyway. Doing something was better than doing nothing.

Reacting to pandemic uncertainties made us covid-ready: hoarding necessities, obsessively cleaning our homes, our hands, even our groceries, wearing gloves and, before long, two masks at a time. We leveraged the traits that uncertainty summons up: caution, focus, planning, attention to detail, and drive.

When we're actively engaged with uncertainty, we're able to get even the smallest details right. It's called a *narrowed scope of attention*. Imagine that you're taking a walk in the woods and happen upon a bear. You freeze, and as it gets closer, your attention narrows to take in every bit of information possible. Does it see me? Is it moving in my direction? Are there any cubs around for it to defend? The danger of the bear is given exponentially greater priority than the aspects of the woods that you were enjoying just moments before: the beautiful trees, the sun-dappled field of wildflowers, the birds sweetly singing. All that disappears as you cope with the danger in front of you. With this narrow scope of attention, you're more likely to survive. Without it, you're likely to just get the gist of the threat—not so helpful if you're trying to avoid a bear mauling.

Now—no need to imagine here—you're living through a global pandemic. The facts remain uncertain, but you need to focus on learning as much about the disease as possible—to take in detailed facts, judge their veracity, update information as needed, and make informed decisions. Can I really catch the virus from surfaces? How important is wearing a mask? What's the evidence that gathering outdoors is safe? The more you learn, the more you give the realistic dangers of the virus attentional priority, while unclear or vague information (which is more likely to be false) fades into the background. This prevents over- or underestimating the threat of the virus and helps you make the best choices possible to stay physically safe and psychologically sound. With this narrow scope of attention, you're more likely to survive.

Narrowing our scope of attention as we gather more information in uncertain times isn't the only way uncertainty helped us during the pandemic. Anya was living in suburban New Jersey when the lockdown went into effect. She and her husband, Mike, were both musicians, and when the pandemic struck, their work lives changed overnight. Anya, who also acted, had no idea when she would be able to return to work and what it would look like. Mike, who had built a successful career on Broadway, was out of work for the foreseeable future.

Uncertainty was nothing new to Anya. As she had built her career, she had become used to hustling from one gig to another, not sure when the next project would come her way—the life of an artist, but one she loved. Prepandemic, she had thought that the key to success was planning well for the future. The pandemic blew that assumption out of the water. Now there was no predicting when the next gig might happen—or if it would happen. How could she plan for the unprecedented unknown? Every day was like running a marathon, one she couldn't train for, and the harder she ran the farther away the finish line seemed.

Then, as the fall approached, she and Mike had to figure out school options for their nine-year-old son. Even after enduring several excruciating five-hour Zoom meetings with the local board of education about school-opening plans, the parents learned that there would be options for after-school sports but little else. As the meeting was ending, one mom spoke up, asking for details about the music program. To her, music education wasn't a luxury; it was an intellectual, emotional, and social necessity. The school superintendent had little to say except "Well, they can't be blowing into instruments during the pandemic. Okay, let's move on."

But that mom was not about to be shoved aside. "I don't really want to move on," she said. "Why don't you have an answer? Why is everything about sports?" She was furious, but she was also worried; it was entirely possible that the kids would lose music education for the year. The uncertainty of it all made her fierce—fiercely protective of the needs of her kids and others. As Anya put it, "there is nothing as powerful as being worried on behalf of your kids. Mess with my kids, and watch out." And you better believe that mom kept at it until the school figured out how to continue the music program.

Uncertainty makes us fierce when we need to be. It also keeps us believing that we can act to gain control over whatever might be coming our way. During the pandemic, I found solace in a supremely mundane control strategy: list making. Don't underestimate the power of a good list. The science of list making—yes, there is a science of list making!—shows that organizing in a linear way what we want to accomplish or remember has numerous benefits. Making a list boosts our sense of well-being and personal control. Studies of memory and aging show that just making a list, particularly one that is well organized and strategic, can help older people remember items and facts as well as younger people do—even without looking at the list.

During lockdown I made written schedules for my children—and myself. They were like signposts, keeping us moving toward a destination, even though we weren't always sure what that destination was. We broke the day down into morning, afternoon, and evening and wrote down activities for each part. At 8:30, Zoom school starts, but there's time for a break before lunch and a nice walk at about 12:30. At 1:00, Zoom school starts again, but luckily it's followed by a family dance party before dinner. Yes! You get the picture.

Those lists gave us a sense of control because they kept

us moving forward with a sense of purpose. And they did even more than that: they sparked new habits. We started hiking as a family and found we loved it, simply because we put it onto the list. We made lists of our favorite recipes and stocked up on the ingredients rather than nuking frozen food every night (although we did that, too). Meals became a welcome ritual that gave us a sense of connection and purpose. I also made lists of the things I wanted to do more often during the pandemic, because embracing uncertainty led to new priorities for me: I was lucky enough to be able to spend more time writing, doing things with my family, and pursuing hobbies I had forgotten about. Others found themselves with less time and new struggles, but whatever our experience of the pandemic, many of us decided to just go for it—because who knew what tomorrow would bring, so what did we have to lose?

This is not to suggest that it was one big list-making kumbaya fest for me and my family. Far from it. There were days full of despair, exhaustion, and hopelessness. My son, who is a worrier, and my daughter, who is not, both struggled with fears of covid and other things. But on those bad days, we went to bed, got up the next morning, and together faced the uncertainty of our lives anew. Maybe we made a list, or maybe we didn't. But we did it together, and every day we tried to take small steps to gain a sense of control, to create certainty out of the uncertainty.

Indeed, the power of togetherness was another lesson we all learned, courtesy of uncertainty. Some of us might believe that raw willpower—suppressing unwanted feelings and actions and resisting short-term temptations to achieve long-term goals—is the best way to overcome adversity. But willpower wasn't enough as we coped with chaos, with such complete and utter disruption of everything we had thought we could count on. We couldn't just will ourselves back to feeling good, to doing the things we needed to do, or to having our normal lives back. And the more we tried, the more—as the science of willpower shows—we felt depleted and ended up less able to control ourselves. Like being on too strict a diet or too intense a workout regimen, eventually we just can't keep it up.

Still, we needed to exercise self-control, caution, and wisdom during the pandemic. So what did we do? If we were lucky, we might have learned what researchers in social psychology have known for almost twenty years: that when we need more self-control and our willpower falls short, our feelings of closeness to, care for, and appreciation of loved ones can fill the gap. Just feeling gratitude toward others, for example, directly improves our self-control. In an adult version of the famous Stanford marshmallow experiment, in which kids are given the choice to eat one marshmallow now or wait to eat two marshmallows later, money is swapped in for marshmallows. Half of the subjects were asked to take a few moments to remember someone they were grateful to, and half were not. Those feeling gratitude were willing to forgo twice as much money now in order to gain more in the future than their ungrateful counterparts were. There again, uncertainty helped by directing them to one of the most precious resources we have: human connection.

What's Anxiety Got to Do with It?

During the pandemic, uncertainty inspired us to take action—everything from wearing masks to making lists, from exercising caution to getting the details right, and from fiercely fighting for what our communities needed to drawing on our satisfying social connections.

But did that make a difference in our anxiety level?

My colleagues and I tracked symptoms of anxiety disorders during the first six months of the covid crisis in 1,339 teens from three countries: the United States, the Netherlands, and Peru. We had already selected those teens to take part in our research prior to the pandemic because they were struggling with severe anxiety. We expected that the pandemic would send them spiraling even further down into debilitating worries and fears.

We were wrong.

Those teens remained resilient; the severity of their anxiety held steady, neither increasing nor decreasing, even as they were forced into lockdown. Research from the United Kingdom showed a similar pattern: anxiety levels remained stable in a group of nineteen thousand eight- to eighteen-year-olds during the pandemic. Not only that; up to 41 percent reported feeling *happier* during lockdown than before the pandemic, and 25 percent reported that their lives were better than in the before times. Although some of these trends are attributable to young people experiencing fewer social demands and stressors (think less peer pressure), lockdown was far from easy.

In other words, the lesson here is not that the overwhelming uncertainty of the pandemic wasn't distressing or anxiety provoking. It definitely was. It's that in the final equation, what determines our well-being is not the existence of overwhelming uncertainty but what we do with it.

Here is where anxiety is the secret sauce. When we feel the tension of the capricious, uncertain future, anxiety animates us to take action. It gives us the courage to avert negative outcomes and sharpens us so that we discover possibilities we might not have imagined before. Anxiety doesn't allow us to sit around passively and become victims. It drives us to do things. And although they might not always be the right things or the effective things, the mere act of doing something—of taking action in response to uncertainty—makes us feel better and in many cases leads to something good. Anxiety isn't the only emotion that can help us achieve this, but it's a powerful one when we learn how to leverage it.

These are the gifts of anxiety. Without it, I believe we wouldn't have persisted through the marathon of the pandemic nearly so well. Think of uncertainty as the starting pistol of the race and anxiety as part of the energy, muscle, and sinew that powered us through to the finish line.

8

Creativity

Thus our human power to resolve the conflict between expectation and reality—our *creative* power—is at the same time our power to transcend neurotic anxiety and to live with normal anxiety.

—Rollo May, *The Meaning of Anxiety*

In 2017, Drew moved to New York City to pursue a career in theater. It was a big transition, so when he was walking around the city one day and started feeling wound up and nervous, it wasn't too surprising. Soon, though, his throat felt tight and dry, making it hard to catch his breath. As it became more difficult to breathe, he felt an oppressive sense of dread, as though something terrible was about to happen. He wandered around for a couple hours, trying to cope. When nothing he did seemed to help, he hopped on the subway to literally and figuratively ride it out. But going underground only made it worse; his heart raced, pains shot through his chest, and he gasped for air. Shaking and sweating, he managed to stumble off the subway and make it home, where he collapsed onto his bed, finally shutting down.

That was Drew's first panic attack. It lasted almost an entire day.

Over the next several months, more attacks followed, and Drew sought help in therapy. It was there that his view of anxiety started to change. "That first panic attack and ones I had after were horrible experiences," he said, "but they were also gifts, because they forced me to finally come face-to-face with my anxiety. As a result, I have grown more as a human being over the past couple years than ever before. Anxiety is my teacher."

Drew didn't shy away from anxiety; he explored it, even creating a multimedia theater piece called *Variations on a Panic Attack*, described as "reimagin[ing] the panicked mind as a commanding melodic-death metal soundscape." In a workshop performance of the piece, Drew and his four-piece band walk onstage, eerie ambient music filling the space. A computerized female voice from the New York City subway cuts through: "This is a World Trade Center–bound E train. The next stop is Fiftieth Street." Speaking into a microphone, Drew describes getting onto the train as panic starts to overwhelm him. The music slowly grows louder and more discordant, until the sounds of the train can be heard, grinding and blaring. It's pretty overwhelming. The audience, increasingly uncomfortable, is unsure how to react. But we are paying attention. Eventually, measure by measure, the cacophonous dissonance becomes the sound of a band playing together in synchrony—still loud, but the melodies and beats are no longer at odds because now they are working together.

Witnessing *Variations on a Panic Attack*, we experienced a piece of what Drew had learned: when we accept the discomfort of anxiety and listen to what it's teaching us, we can grow and create and in the end resolve the inner dissonance we feel when we're anxious. Some acts of creation inspired by anxiety, such as *Variations*, are works of art. Others are so simple and mundane that they don't seem creative in the least—until, that is, we get to the heart of creativity.

On Creativity and Wilted Cauliflower

When we think of creativity, we default to thinking about artistic endeavors: a painting, a book, a musical performance. We might also include inventors who create new technologies or better widgets. But that's narrow thinking. Creativity is what *all* humans do, and we do it constantly.

That's because creativity is any transformation we make from one state to another. It's creativity when, in one moment, our mind is a blank, and in the next there's an idea there; when we make something new, something that has never existed in quite the same way before—even if it's a ham sandwich. It's when we generate ideas or recognize a good idea when we see it. It's coming up with alternatives when a solution isn't working so that we can figure out a problem and communicate it. Creativity is seeing connections where others might not and pursuing them with curiosity, energy, and openness. Take Blockbuster and combine it with Amazon, and you get Netflix.

Creativity is seeing possibilities.

It's the end of the workday, and I'm absorbed with deadlines and catching up on all those unanswered two-week-old emails that hang like an albatross around my neck. I look at the clock and—oh, no!—it's dinnertime. I haven't even thought about what to make my kids. I run downstairs to choruses of "I'm hungry! What's for dinner? Can I have a snack?" I open the fridge, and it's empty. Some cheese, milk, eggs (expired?), and in the veggie crisper, a head of slightly wilted cauliflower. My heart sinks and then starts to race as I get that feeling in the pit of my stomach that may or may not be a burst of adrenaline. What to do? I could just order a pizza—for the third time this week. (And it's only Wednesday.) But I want my kids to have healthy meals more nights than not, so I take a deep breath and start thinking. An old cauliflower isn't a great start, but wait, there's such a thing as the internet—so I search "dinner with leftover cauliflower." Literally the first thing that comes up is an article entitled "Thirteen Ways to Use up Leftover Cauliflower." Thirteen! Now my only problem is to choose among this embarrassment of riches—should I make the roasted cauliflower cheese casserole or the keto cauliflower fritters? Thirty minutes later, I've taken that sad, forgotten cauliflower and created a new dinner favorite.

It wasn't my relaxed attitude about mealtimes or my laissez-faire parenting style that spurred me to create a healthy dinner, it was anxiety: anxiety about whether my kids were eating well, anxiety from being caught by surprise and unprepared for dinner, anxiety because I care about coming together over a hot meal that doesn't come from a box. Our lives are littered with anxious moments, big and small, that make us more creative because they help us see the possibilities—even in cauliflower—so that we can bring something worthwhile into being that has never existed before.

Creativity is seeing possibilities. And anxiety helps us see the possibility that there are possibilities.

Anxiety also influences *how* we're creative—what researchers call fluency, or the sheer number of ideas or insights someone comes up with, and originality, how novel those ideas are. Both of these aspects of creativity change along with our moods.

How do we know this? Researchers first induce specific moods in their subjects—by asking them to write an essay about a situation that provoked strong emotions or watch emotionally intense movie scenes. Then they measure the subjects' creativity. It turns out that it isn't whether our moods are positive or negative that influences our creativity, it's whether they're activating or deactivating—in other words, whether they *move* us. Activating moods such as anger, joy, and anxiety increase our energy level and motivate us to do *something*. Although they're a mix of positive and negative feelings, these activating emotions are a breed apart from deactivating emotions such as sadness, depression, relaxation, and serenity, because the latter just slow us down.

A 2008 study conducted by researchers in Europe and Israel induced activating and deactivating moods in participants, then asked them to do something creative—to brainstorm ways to improve the quality of teaching in the university psychology department, jotting down as many ideas, solutions, and suggestions as they could think of. Deactivating moods didn't have any effect on creativity, but activating moods, regardless of whether they were positive or negative, prompted more fluency and originality. People who felt moderately more anxious (also angry or joyful) came up with more ideas, and ones that were more innovative. One reason anxiety in particular increased creativity was that it prompted people to stick with brainstorming and problem solving longer. They persisted.

Activating emotions such as anxiety not only help us persist but can help us balance out deactivating emotions that can disrupt our creativity. If anxiety inspires us to see the possibility of possibilities and persist in creative efforts, even when emotional distress slows us down, what about the times when anxiety itself is a burden? Is it a wellspring of creativity then?

The other week, I woke up in the middle of the night, heart racing and sweating, with feelings of dread—not so different from what Drew had experienced while walking the New York City streets. It was 3:17 a.m. Immediately, my mind was drawn to worries about my relationship with a close colleague. She and I were at odds over—well, pretty much everything, it seemed. Thoughts wouldn't stop running through my head, like being on a treadmill, and I kept remembering all the things I was upset about, as well as the last frustrating conversation I'd had with her and what I *should* have said instead of the totally inadequate things I had said. I don't have to interpret these feelings for you. This was anxiety. What's creative about that?

This kind of distressing anxiety *is* creative because it's a call—a summons to listen and heed the smoke alarm going off, telling us that there might be a fire. It's a call to dig deeper into what's going on in our hearts and minds, rather than just skim above it as we normally do because we fear being dragged down into our emotions.

I decided to listen to that anxious voice, so when I finally got out of bed, after tossing and turning for a few hours, I knew what I had to do: I needed to start a conversation with my colleague, and it needed to be honest. Just making that decision cleared away the fog of the night's worries. And it reminded me that I had a good bit of control over the situation—and could make it better by doing more than just tossing and turning in bed at night.

Anxiety is a wellspring of creativity *because* it is uncomfortable. If we allow ourselves to experience discomfort, then we want to resolve it. *We need to.* So we take actions that will make our lives better and create the future we want. Turning our backs on anxiety means turning our backs on possibilities.

When our response to anxiety is to become creative— when we paint, plant a beautiful garden, start a difficult conversation, or take an old piece of cauliflower in the fridge and turn it into a pretty decent meal—we can see that positive choices, not dread and fear, are the gifts of anxiety.

We can use anxiety to see possibilities creatively and persist in turning them into reality. But even with that there's a risk. It's called perfectionism—and it, too, can sometimes be inspired by anxiety.

Never Mind Perfectionism, Here's Excellencism

Anxiety and perfectionism have some things in common—like anxiety, perfectionism keeps us caring about what happens in the future and energizes us to get it right. In that sense, it's a great stimulus if we want to achieve and create—until it isn't. And unfortunately, it isn't great a whole lot of the time, because perfectionism isn't just caring about what happens in the future and striving with high standards to get it right; it's about what happens when we fail.

The standards to which perfectionists hold themselves are self-explanatory—they're unrealistic, overly demanding, and often impossible to achieve. What happens when perfectionists fail to achieve perfection? They don't bounce back and move on or take pride in improving on their personal best. And they definitely don't celebrate small accomplishments along the way. Instead, they beat themselves up with harsh self-criticism. To a perfectionist, life is all or nothing: you can be a winner or you can be an abject, worthless failure, with nothing in between. This relentless pursuit of flawlessness inexorably leads to low self-esteem, depression, and fear of failure. As a result, perfectionists often end up achieving much less than they aspire to because they hold back, procrastinate, and even stop taking on challenges altogether—because it's better to not have entered the race at all than to have spun out in ignominy.

Though there are similarities between anxiety and perfectionism, whereas anxiety keeps us moving forward, trying to figure out solutions when we hit barriers, trying to make good things happen, perfectionism stops us. Leaving no room for failure or uncertainty, perfectionism narrows our path until we can no longer advance. Perfectionism—like extreme, unhealthy anxiety—shuts down possibilities.

Luckily, there is an alternative to perfectionism that draws on healthy anxiety but increases our ability to persist and create. It's called *excellencism*—working toward excellence rather than perfection. It involves setting high standards but *not* beating ourselves up when we don't meet them. An excellencist is open to new experiences, takes unique approaches to problem solving, and is okay with getting it wrong—as long as they can learn from their mistakes to strive toward exceptional achievement.

Excellencists often show higher levels of anxiety compared to nonperfectionistic people—along with greater conscientiousness, higher intrinsic motivation, enhanced ability to make progress on goals, and more feelings of positive well-being. What they don't show is more debilitating anxiety. They also don't tend to carry other burdens of perfectionism:

higher rates of burnout, intense procrastination, long-term depression, and suicidality.

Excellencism takes the best part of perfectionism—caring about getting details right, putting our hearts and souls into what we're creating or accomplishing—but opens up rather than shuts down what we can achieve. The rate-of-returns analogy shows us how.

Most of us assume that hard work pays off and, conversely, that if a task should take a day and we devote only an hour to it, our results won't be up to snuff. Research consistently backs up this intuition; when students invest more time, effort, and energy into studying, their grades go up. When people set themselves difficult goals, they generally outperform those with easy goals because they put in extra effort and personal investment. As the input of time and energy increases, the output of success and performance rises proportionally. This is the zone of increasing returns; one unit of work pays off in one unit of improvement. It's simple math.

The math, however, isn't quite that simple, because it turns out that it's not just quantity of effort that matters; quality does, too. The more purposeful we are in our efforts and the more clear and attainable our goals, the better our performance and learning. Simple quantity of effort can backfire. And when it does, we hit the point of diminishing returns; efficiency flies out the window, and putting in more time and effort yields smaller and smaller improvements. Worse yet, diminishing returns can escalate into decreasing returns, where putting in more time and effort actually makes things worse. It's like adding extra hours of training at the gym on top of the recommended regimen, only to realize that you've overtrained and are so depleted that you can't even do the basics anymore. Or continuing to tweeze your eyebrows in pursuit of that perfectly shaped arch until your brow has all but disappeared—and you have to draw it in with a pencil, as your grandmother did. That's where perfectionism tends to land us—in the zones of diminishing and decreasing returns, where more effort to achieve elusive perfection just makes us less productive and less creative. And with slender eyebrows.

We can break down any task into zones of increasing, diminishing, and decreasing returns. Imagine that two people—one a perfectionist and one an excellencist—are writing a short story. In which zone will each land? Both have to figure out how much time they need to spend: too little, and the plot will be muddled, the writing disorganized, and the grammar atrocious; just enough, and they'll be in the zone of increasing returns—the quality of the story will proportionally improve with each hour of effort. It's when they're close to being finished that the differences between perfectionists and excellencists really stand out. Perfectionists are much more likely to enter the zone of diminishing returns, in which each hour of labor yields smaller and smaller improvements in organization, clarity, and creativity.

That's why, whether when writing a story or doing something that's perhaps a bit more boring, such as proofreading, perfectionists, counterintuitively, turn out lower-quality work than they're capable of doing. For example, research shows that perfectionists take longer than nonperfectionists to do repetitive or boring tasks, create more inaccuracies, and work less efficiently. An obsession with flawlessness affects scientists in much the same way: highly perfectionistic scientists create lower-quality, less creative, and fewer published papers.

Excellencists, on the other hand, tend to sidestep these danger zones. They find the sweet spot between the perfect and the merely okay—because they can be excellent without being perfect. They operate within the zone of increasing returns for longer periods of time because they aim at high but attainable standards and invest sufficient—but not excessive—effort to reach their personal best. And they know when to give it a break. They're not stuck on the exhausting treadmill of perfection.

Not only does excellencism help people be more efficient and productive, it improves the quality of what they create. In a 2012 study, almost two thousand undergraduates were rated on the degree to which they showed excellencism—having personal standards that were high but left room for making mistakes. Then they completed standardized tasks probing multiple levels of creativity—from coming up with a witty caption for a cartoon to the more challenging task of generating original, high-quality solutions to real-world conflicts. The degree to which someone was an excellencist predicted the quality of his or her solutions in the more challenging—but not less challenging—creativity tasks. In other words, the greater the excellencism, the higher the quality of the solutions. Excellencism does actually make people *more* excellent than perfectionism does when it really matters.

Thomas Edison said, "I have not failed. I have just found ten thousand ways that won't work." This is excellencism—fueled by anxiety—in action. It's the capacity to see that when one possibility shuts down because it failed, another opens up and allows us to strive toward even greater, more creative achievements.

Future Calling

Many people heed the call of anxiety and use it to reach their goals. Among their great strengths—in some cases their genius—is that they anticipate the unknowable, uncertain future and find ways to move beyond their comfort zones to envision and create something that hasn't been achieved before. Even when their anxiety gets tough, almost as though they're drowning in it, they jump into the river and start swimming—into the future.

Massively successful tech entrepreneurs are an example. Putting aside the many, many things we can critique about them, their undeniable achievements show a common uncompromising focus on the future. Take the 2021 billionaire space race, in which Richard Branson, Jeff Bezos, and Elon Musk competed to be the first rocket company owner to orbit the earth. If they did so to inspire the masses during the pandemic, it completely backfired; most people saw their flights as flashy joy rides for the obscenely rich. But it shows that although some people look into the future and want to see more of the status quo, others see possibility—and it drives them. Elon Musk in particular has focused his energies on shaping the future. It looks like science fiction—sending humans to Mars, creating implantable brain-computer interfaces, and preventing evil AIs from taking over the world. Whatever our opinion of Musk or any of the other entrepreneurs, one thing is beyond debate: they've pushed the boundaries of what is possible in the present to create the future they want to see—for better or for worse. Whatever's causing their anxieties today, their attention, their efforts, and quite a bit of their fortunes are focused on the future.

Bringing the subject back down to earth, people use anxiety all the time to make choices about the future. These choices may be less grand than space travel and brain-computer interfaces, but they have the potential to have a truly positive impact on people's lives. A study by researchers at the University of Alabama examined the characteristics of people who reliably pursue the follow-up care they need after heart transplant surgery. Participation in follow-up care is a powerful predictor of recovery and prognosis, but a large percentage of patients pursue only some of the recommended procedures and assessments, and some never do any follow-up at all. Anecdotally, medical professionals know that anxiety over health is what stops most people from adhering to treatment; they're so worried that they will learn they're not doing well that they avoid going to the doctor altogether. But burying our heads in the sand isn't the best strategy. We have to tolerate and manage the anxiety of an uncertain prognosis and persist in treatment. It might even spur us on to make extra effort to take good care of ourselves. And that's just what the researchers found. People with some anxiety, but not extreme levels, were more likely to receive the recommended treatment—*and survive*—after transplant surgery. In this case, using anxiety to make future decisions just might have saved their lives.

Anxiety Is Freedom

If uncertainty is the starting pistol and anxiety the energy that helps us persist to the finish line, creativity is the race itself, full of possibilities. In other words, creativity arises in the gap between present reality and future possibilities. That's also where we experience the discomfort of anxiety, and if we can tolerate it—listen to what it's telling us—we can plan for the future, envision works of art, and hatch new ideas. We don't create something wonderful lying on the couch, catching a few z's. We create something excellent through struggling and throwing ourselves into the gap. If the gap is extreme, we feel conflict and distress. If there's no conflict, we live without momentum and are stuck standing still. Life is a series of such gaps, of different sizes.

Earlier, when I described Dr. Scott Parazynski's remarkable 2007 space walk to repair the International Space

Station, I didn't mention that the veteran of five space shuttle flights wasn't cool as a cucumber in every situation. In fact, although a true adventurer—one of the few people who have reached the summits of both space and Mount Everest—Scott was terrified of what some might say is the inverse of reaching those great heights, cave spelunking. Going down into the deep, dark bowels of the earth made him feel claustrophobic. It was his personal challenge, his big gap, where he felt the most intense, uncomfortable anxiety.

That's how it is with people who use anxiety creatively and well; they don't love their anxiety or necessarily have total mastery over it in every situation. And that's okay. Because in some key, important domains of their lives, they feel in anxiety—to quote Kierkegaard—the dizziness of freedom, because it helps them bring something new into existence. They feel in anxiety creative, boundless possibility. And they lean toward it rather than away.

9

Kids Are Not Fragile

If an anxiety, like light and the shadows of clouds, passes over your hands and over all you do, you must suppose that something is acting upon you, that life has not forgotten you, that it holds you in its hands. It will not let you fall.

—Rainer Maria Rilke, *Letters to a Young Poet*

When my son was nine, I decided it was high time he learned to ride a bike. He's a city kid, so although he had been zooming around downtown Manhattan on a scooter since he was four, bikes were a mystery to him. That bothered me. Was he missing out on an ideal childhood? Would he be left behind when his friends tore off on their bikes, Goonies-like, to pursue some adventure? So that summer, while spending time in upstate New York, I knew it was the crucial moment to teach him. I had an old BMX from the 1980s—the Gremlin—stashed in the garage. Believe me when I tell you, they don't make 'em like that anymore.

Compared to the ultralight bikes kids use today, it was a beast—solid, heavy, almost tanklike; not easy to learn on.

Despite the challenge of the Gremlin, Kavi did pretty well his first time out—but he was miserable. He kept complaining that it was hard and he was exhausted. He started whining "What if I fall?" and finally, in a half whisper, "I'm scared." But I was not to be deterred. I minimized— "Oh, what's the worst that can happen, a skinned knee?"; I pushed—"Come on, buddy, stay focused! Look in front of you!"; I exuded bonhomie—"You can do this, honey! Come on, you're great. You're *amazing*!" After thirty minutes of that, he looked like a tight little ball of stress, so I called it quits. As we trudged up the hill to the house, I continued to mete out what I was sure were helpful insights and sage advice.

Once we got home, Kavi tore off to his room without a word. Sighing, I took my phone, which I had used to film Kavi biking, out of my pocket and saw that it was still recording. It must have captured everything from the lesson to the pep talk while we walked up the hill. Great, I thought, I'll listen and maybe figure out when and how things went sideways.

If I'd known what I was about to hear, I would have deleted it right then.

ME: *Fine. All right. Let's go, Kavi. I'm ready to give up. I'm doing my best to be supportive, and you're being nothing but a grouch.*
KAVI: *(tears at the edge of his voice) I'm trying.*

ME: *You're doing great. You're just complaining every second of the way.*

KAVI: *I'm really trying.*

ME: *You're amazing. Why are so you negative about it?*

KAVI: *I don't know. I'm scared.*

ME: *You are not scared. There is nothing to be scared about. You did it perfectly. You didn't fall down once. Maybe I should knock you off the bike so you can see it's not a big deal, and then you'd get over it.*

KAVI: *(whimpering sound)*

ME: *Kavi, honestly, you're just talking yourself into being scared. And I don't know why.*

KAVI: *You're right.*

ME: *You're crushing it. You're so good at it. And you're just talking yourself into "I'm scared, I'm scared." No, you're not. You're doing awesome. You haven't even fallen, haven't even gotten a single bruise.*

KAVI: *I know.*

ME: *So I have to give you a little tough love here. You gotta get your head together.*

It went on like that for another minute or so.

When it was over, I was in tears. My perception of what I had said was completely different from reality. Rather than being tough but supportive, I had been a caricature of a type-A parent—discounting his feelings, shaming him, demanding that he perform, and essentially telling him to "man up." I knew I had to fix that, but I was troubled by the question of why. Why had I run roughshod over his understandable anxieties about learning to ride a bike? I intellectually knew better, so it meant that there was only one answer: I wanted his anxiety to go away because it was making *me* uncomfortable. Why? Because it meant he was fragile.

Luckily, nothing could have been further from the truth.

Antifragility

What is a parent's job? When our children are very young, it's to protect them, fix things that go wrong, and make sure they have food in their bellies. As they grow and enter adolescence, we shift into a consultant role, where we support, advise, and teach them the skills they need to fix things on their own. As a consultant, when my son gets into a fight with his friends, I'll start by brainstorming with him about how to handle it himself rather than calling up his friends' parents to intervene. When my daughter gets a bad grade, I'll first talk with her about steps she can take to study more effectively and seek support from her teacher instead of calling up the teacher myself to kvetch and question the grade. As our kids grow up, our job is to do less rather than more, to give them a chance to fall down and pick themselves back up again.

But as we look around at kids today, sobering statistics might make us question the wisdom of merely being a consultant—there are so many ways they can fall, and can they really pick themselves back up?

Let's start with our kids' futures. Putting aside disastrous climate change, the threat of future pandemics, and disturbing political trends, can we at least assure them what many of us were guaranteed growing up: work hard and you'll have a good life? Maybe not. Young people today, compared to their parents and grandparents, are less likely to be gainfully employed, are less likely to own a home, have only a fifty-fifty chance that they will outearn past generations at the same age, and are more likely to be shouldering heavier student loan burdens.

Then there's our kids' mental health; they are struggling. Parents and schools have raised the alarm that worries and fears are getting into the way of even very young children's ability to learn, get along with others, and have fun—kids being kids. And once they approach adolescence, the situation gets more concerning. In any given year, 18 percent of teenagers will suffer from an anxiety disorder, and by the time they turn eighteen, a staggering 33 percent of them will have done so—that's more than 10 million in the United States alone. The kids are fully aware of the scope of the problem. A Pew Research Center Report released in February 2019 showed that 96 percent of teens surveyed believed that anxiety and depression were a significant problem among their peers, with 70 percent saying it was a major problem. And they're correct, because the 10 million young people who experience an anxiety disorder are also much more likely to suffer in adulthood not only from continued anxiety but also from depression, addiction, and medical problems. Teen anxiety is a gateway to future poor mental health and illness.

We take these statistics as signs of this generation's fragility. It helps explains the rise of the trend I discussed in chapter 4: the proliferation of safe spaces and trigger warnings. The point here isn't "Don't worry about the future." Frankly, I do, too. But constantly shielding our kids from emotional distress—and teaching them to do the same—is not the solution. It's the opposite of what we should do because, despite the stresses we face in the world, we humans are not fragile. We're antifragile.

Something that is fragile breaks easily and should be handled gingerly. When something fragile does break—picture a china teacup falling out of your hands and smashing into little pieces on the ground—it can never be put back together as it was because the cracks will always show.

Antifragility is the opposite of fragility. It's the quality of growing stronger *because of* challenges, difficulties, and uncertainties. That makes it distinct from related concepts such as resilience, robustness, and the ability to resist and bounce back from a challenge. Antifragile things don't just bounce back like a flexible branch that doesn't snap in a storm; they actually gain from randomness, volatility, and disorder. They need chaos to flourish.

That's why humans are fundamentally antifragile.

Take the immune system. It is antifragile because it requires exposure to germs and pathogens that challenge it so that it can learn to mount immune responses. Without such exposure, we are like the boy in the plastic bubble who, lacking a functioning immune system, can't survive in open spaces. Indeed, when there are no challenges to overcome, systems that are antifragile become rigid, weak, and inefficient. When life is always predictable, safe, and comfortable, there is no need to respond with effort and creativity. Bones and muscles are antifragile for that reason: spending a month in bed leads to atrophy; challenging our bodies makes them stronger.

Anxiety is antifragile, too. When we allow ourselves to feel the discomfort of our worries, fears, and uncertainties, we are challenged—but we are also motivated to take actions that overcome problems and ease the pain. As a result, we will manage anxiety better the next time. When we make a huge mistake, it is our ability to suffer through the anxiety of getting it all wrong that strengthens our ability to persist the next time we blow it. In other words, the way to grow a strong emotional immune system is to allow ourselves to feel difficult emotions and push ourselves to bear emotional pain. If we construct our lives with the goal of avoiding these unhappy feelings and of destroying all forms of uncertainty and randomness, we will be prevented from using our antifragile natures to navigate the challenges of life to the best of our ability.

From this perspective, protecting our children from their anxiety is exactly the wrong thing to do. Without the opportunity to put anxiety into practice, children can't learn to find possibility in uncertainty and to be creative in the face of adversity. We aren't born knowing how to manage anxiety, just as we aren't born with an immune system that has optimized its germ-fighting power. But both of these antifragile systems learn from challenge and are capable of finding their way.

Nassim Nicholas Taleb, who coined the term *antifragility*, described it beautifully in his book on the subject: wind extinguishes a candle but energizes a fire, he wrote, and so "you want to be the fire and wish for the wind."

That's not to say we should let our kids face overwhelming challenges alone and unaided; even a blazing fire of antifragility can be extinguished by a hurricane of emotional stress or trauma. Allowing them to be in high-intensity, high-challenge situations needs to be balanced out with providing them with comfort and support. But research confirms that when it comes to anxiety, we have to let our children feel the wind.

In 2019, 124 kids between ages seven and fourteen and their parents went to the Yale School of Medicine Child Study Center to take part in a research study with their children, all of whom were diagnosed with an anxiety disorder. They went for gold-standard cognitive behavioral therapy (CBT) because they knew that it was the most tried and true treatment for anxiety. Their kids would explore their worries and fears, gradually learn to confront them, identify and revise unhelpful thoughts such as catastrophizing and harsh self-criticism, and try out new strategies and behaviors to handle their anxiety. Yet as part of the study, half of the parents agreed to forgo that top-notch therapy for their children and instead enrolled in therapy themselves. It was a new type of parenting therapy, and it had one very specific goal: to teach parents to stop taking away their children's anxiety.

The parenting therapy was called SPACE—Supportive Parenting for Anxious Childhood Emotions. It was laser focused on the fact that parents of anxious children tend to overaccommodate their offsprings' anxieties. If a child has a fear of flying, the family takes a driving vacation instead; if a child is shy and socially anxious, the parents stop entertaining friends at home; and if a child can't stand to be separated from the family, they'll spend every possible moment with the child, even allowing him or her to skip school. Such well-intentioned efforts are meant to help their children, but as any parent will tell you, it also helps the parent—it's hard to see our kids suffer and struggle, and by comforting them, we comfort ourselves.

But that tactic usually backfires. Avoiding anxiety-provoking situations may comfort our anxious child in the moment, but in the long run this accommodation prevents kids from learning to handle situations that cause anxiety.

SPACE taught parents to allow their children to be anxious but to do so in a supportive way—acknowledging their children's emotions, conveying confidence in their ability to cope, and helping them through rather than around anxious situations. For example, if Solveig refuses to go to school because she can't bear to be separated from her father, Dad is taught to say, "I know you feel upset right now, but you can handle it. You'll be all right"—and then send Solveig to school anyway. If Kabir's family has stopped allowing visitors at home because he suffers from extreme shyness and won't leave his room when others are in the house, his parents invite trusted friends and family members over, for short visits at first and then for longer ones, and make sure Kabir mingles so he can slowly grow more comfortable and confident.

Change doesn't happen overnight, but after twelve weeks of parenting therapy, 87 percent of the children whose parents received SPACE therapy showed significantly less severe anxiety and more positive adjustment—outcomes as good as for the children who received top-shelf CBT. By accommodating their kids less and supporting them more, those parents not only helped their kids but learned that their children weren't as fragile as they had feared.

We're not all lucky enough to attend parenting therapy. But there are small things we can do to promote our kids' antifragility for anxiety, like booster shots to their emotional immune systems.

For one, we can allow our children to sit with their anxious feelings. When my son was in fourth grade, he left his math homework at school one day. When he realized that it wasn't in his bag, he burst into tears, paced around, and started gasping for air as though he were hyperventilating. I gave him a glass of water and sat him down. We came up with a good solution together: we'd ask a classmate's mom to send a picture of the homework so Kavi could copy it out by hand.

Problem solved! But not quite, because then he revealed what really worried him: that his beloved teacher, Ms. Z., would still know that he had forgotten his homework and think badly of him. The anticipation of having to face her disapproval in the morning ratcheted up his anxiety tenfold.

He begged me to email Ms. Z. to let her know that, in fact, he did do his homework and would be dutifully turning it in in the morning. Just talking about my emailing her visibly calmed him down.

But to his dismay, I refused. I explained why—how working through uncomfortable anxiety is how we learn to cope with it. He wasn't buying it. So now he was worried *and* superannoyed with me. Soon I was feeling my own anxiety level rise; it wasn't easy seeing him so upset over something I refused to do for him. We even did a little mini-CBT session that helped both of us feel better—things such as exploring the details of his worries, talking through whether Ms. Z. really would be upset with him, practicing calming breathing exercises. And even though he became calmer, his anxiety wasn't gone—when he went to bed that night, he was restless and worried.

The next day when he got home from school, he ran up to me, waving a paper in his hand: "I won't tell you. I'm going to show you!" There was his handwritten math homework with a big A+ plastered at the top, along with the words "Wonderful job figuring out how to get your homework done!" Kavi had discovered that the rewards of finding a creative solution sometimes go hand in hand with the anxiety of trying something new.

I could have emailed Ms. Z. and taken his anxiety away. I would have done so with the best of intentions, and he—and I—would have had a better night's sleep. But he would have missed an opportunity to learn that he could tolerate the discomfort of anxiety—and achieve something positive in the process. It is in such commonplace, unremarkable moments that we support or sabotage emotional antifragility. Unfortunately, unintentional sabotage is fast becoming the new normal.

Emotional Snowplowing

Protective parenting has evolved over the past fifty years. In the 1970s and '80s, the idea of stranger danger and the missing child movement took hold, accelerated by the tragic disappearance of six-year-old Etan Patz, the first "milk carton child," from downtown Manhattan in 1979. Over the next two decades, letting kids play without adults outdoors and in public places seemed risky, so that by the late 1990s and early 2000s, kids were allowed 50 percent less time in unsupervised and unstructured play and recess compared to the 1970s. By then, moms and dads had internalized the belief that their children had to be constantly monitored and managed. They became "helicopter parents," hovering over every aspect of their children's lives from education and sports to friendships and fun.

Now, in the twenty-first century, we see the apotheosis of helicopter parents—snowplow parents—forcibly removing every potential obstacle from their children's paths. They are, as the saying goes, preparing the road for the child instead of preparing the child for the road—even if that means breaking the law.

Take a particularly egregious example, the 2019 college

admissions scandal. Dozens of rich and famous parents cheated their children's way into top colleges. It came out that they had paid college athletic coaches hundreds of thousands of dollars to recruit their children for sports they had never played, even staging fake photo shoots complete with uniforms, gear, and trophies for water polo, sailing, and rowing. They bribed examiners to falsify SAT scores. They paid psychologists to diagnose learning disabilities to earn extra time for the ACT exam.

But this extreme example belies the fact that snowplow parenting isn't always about removing concrete, external obstacles to success. It's also about removing internal obstacles—emotions such as anxiety that we think make our children vulnerable and less likely to succeed. Think of it as emotional snowplowing.

I was performing emotional snowplowing when I tried teaching Kavi to ride a bike. My reaction to his fears and anxieties was that *they* were the obstacles to doing what he *should* be able to do—which, in my mind's eye, would be to hop blithely onto the bike, wobble a bit, but then ride like the wind. His anxiety destroyed my dream, and so I just wanted the anxiety out of the way. It also unnerved me that he was struggling when I had thought it would be easy for him. Was he becoming an "anxious child"? Was it a sign that he'd be frightened of future challenges? I couldn't see that it was reasonable for him to feel anxious—about falling down, skinning his knee on the gravel, going too fast on the steep road I'd forced him to learn on. I also couldn't see that my efforts to snowplow through his anxiety were creating a new thing for him to be anxious about—that he would let me down.

Even our well-intentioned efforts to help our kids when they struggle with anxiety can end up being an exercise in snowplowing.

In April 2019, I gave a talk on child anxiety to a room full of parents whose children attended a top gifted and talented high school in Manhattan. Kids must have IQ scores in the stratosphere to get into the school and then must maintain outstanding grades and participate in a slew of impressive extracurriculars. So when a dozen parents came up to me afterward, I figured I would hear stories that are fairly common about gifted kids feeling stressed and worried about the high academic expectations. But the parents described kids who were way past that point. They were falling apart at age fifteen, so overwhelmed by schoolwork that they could barely sleep or eat, constantly self-critical ("I'm stupid, I don't even belong in this school"), and suffering from such crippling anxiety that they froze during tests even though they had mastered the material.

Despite the fact that those parents came to a talk about child anxiety and clearly cared for and worried about their kids, almost no one asked me about anxiety—or therapy or even child emotional development. Instead, they asked my thoughts on how much tutoring was too much, the minimum amount of sleep teens need, and whether competitive sports help kids develop more grit. One father put it this way: "I don't like to push my son to do math tutoring twice a week, chess, and computer programming, but if it helps him measure up to his classmates, maybe he'll feel less stressed."

Their kids' anxiety had spiraled out of control, but the parents didn't want anxiety to be *the* problem. I could understand why they believed that would mean their kids were fragile, liable to be shattered beyond repair. It was like my own mindset when I had pressured my son to ride that bike. To me, and to those parents, anxiety was a disability rather than what it really is—something to explore, discuss, and struggle through. Something to heed. And more important, something that helps their children move forward.

The Marvelous Teenage Brain

Joseph, a sophomore in college, has been busy. As a freshman, he started a nonprofit organization that cleans up plastic debris littering the oceans, and this year he has lent his programming skills to improve the crisis text line on his campus. When asked what he plans to do next, he has a laundry list of possibilities, from throwing a surprise birthday party for his boyfriend to founding a technology start-up. Yet despite his brilliance and ambition, he seems to share the view that most people have about the teenage brain: "I took neuroscience classes and know my frontal lobes are still developing, so I don't always trust my decisions when I'm upset or under pressure."

Joseph is unknowingly parroting a narrative that has infiltrated our view of adolescence—that teens are overly emotional, impulsive risk takers because their frontal lobes are too immature to control their drives and passions. Combine this with older ideas about "raging hormones," and we must conclude that youth is a fragile time of inevitable *Sturm und Drang* when feelings always triumph over logic.

Yet far from being immature and out of control, the teen brain develops in a way that confers much more advantage than we give it credit for.

Just a few years ago, scientists assumed that major changes in the structure and function of the brain were limited to the prenatal period and the first few years of life. We now know that we were wrong and that massive and fundamental development and reorganization continue throughout adolescence and into early adulthood—that is, between the ages of twelve and twenty-five. This means that it's only in our midtwenties that our brains mature into adulthood. But what does it mean to have a mature brain?

The brain develops due to changes in its gray matter and white matter. Gray matter is made up of brain cells and the synaptic connections among them, and white matter is composed of the axons that enable neurons in the outer layers of the brain, such as the prefrontal cortex, to rapidly communicate with those in the deeper regions, such as the limbic system. As a brain matures, the gray matter should thin out as the white matter increases. That's because neural circuits are created and refined through pruning, wherein unused connections between the brain cells—the gray matter—are destroyed, increasing the strength of efficient and useful neural circuits that do the things we want them to do.

It's use it or lose it. As when I learned a bit of Italian in high school but then never studied the language again, the Italian-speaking connections I made were gradually pruned away, so that by now, I can say only *Grazie mille* and *Prego*. It's like pruning dead branches off a tree so it can grow better or deleting old apps from your phone so it will run faster. These aren't just metaphors. A 2006 study published in *Nature* revealed that children with higher IQs showed early growth in gray matter, followed by vigorous thinning of it by early adolescence.

In the human brain, the first areas to mature are the sensory and motor systems that support the five senses and the coordination of body movements. Next in the developmental queue are the limbic and reward systems—the "emotional centers"—of the brain. The last areas to mature are parts of the prefrontal cortex, the "control centers" of the brain, which help us plan, make reasoned decisions, assess risk, delay gratification, and regulate our emotions. How should this imbalance in the development of the emotional and control centers of the adolescent brain be interpreted? Usually it's some version of "Poor teens! They're stuck thinking with their 'emotional brains,' while we adults can think with our 'rational brains.'"

Far from it. Despite this unequal development, the balance of power between the frontal lobes and the limbic system is in constant flux. Sometimes the "control centers" are in the driver's seat; teens can make perfectly rational plans and decisions, follow the rules, and avoid danger. Other times the "emotional centers" are more in charge, and teens prioritize the three R's—risk, reward, and relationships—more than the average adult does. This means that they react more intensely and frequently to emotional information in the world: threats and rewards, love and hate, uncertainty and novelty. But this flux is a double-edged sword. It's advantageous when it allows teens to be flexible, adapting quickly to changes, learning quickly, and tuning in to social and emotional signals around them. But it can also get in the way.

Risk taking is a good example. Because of the imbalance between the emotional and control centers of the brain, teens do indeed take more risks than adults—and even more than children whose prefrontal cortices are less developed. But we see these risky behaviors only under certain circumstances. One of them involves other people. In a 2005 study, young teens (ages thirteen to sixteen), older teens and young adults (eighteen to twenty-two), and adults (over twenty-four) did a driving simulation in which they were told to drive as far as possible until a traffic light turned red and a wall appeared. If they waited too long to stop, they would crash into the wall and lose points. Some of them did the simulation alone, while others were in groups of three people their own age. Guess who crashed the most? The young teens—but only when they were with their peers. The adults drove the same whether they were alone or with other adults.

Through the lens of evolutionary theory, this "problem" of risk taking among young teens with their peers isn't so problematic. Indeed, being open to risk and social connections was an invaluable asset to prehistoric humans, who were, for all intents and purposes, adults by the time they reached early adolescence. That is, once humans were old enough to procreate, they did so; they left the security of the family to form new families of their own, held serious responsibilities that benefited the tribe, and went out into the world to explore and learn. Most ancient humans would have died by age forty, so without adolescent risky behavior, the tribe would have had a serious population deficit in talent—a brain drain, if you will—when it came to doing what it took to survive and thrive. Who would explore new lands and meet new people? Who would lead the dangerous hunting or gathering mission? Who would figure out that fire both created and destroyed and teach other people how to harness it? The adult brain—with its diminished taste for risk and reward and its slower adaptation to change—is not nearly so well suited to pursue such goals compared to the fleet teen brain.

It's interesting to consider that this staggered brain development radically differs from that of nonhuman primates. For example, like humans, rhesus monkeys and chimpanzees are born with immature brains. But unlike in humans, all the cortical areas of these primates' brains mature at the same rate. Evolutionary biologists will tell you that this divergence from our primate cousins must confer an advantage—and support something about us that is uniquely human.

Yes, the marvelous teenage brain isn't perfect. It might even be best suited to the life of a prehistoric human, when teens were independent adults. But where there is risk, there is also opportunity. Teen brains aren't aberrant or irrational; they are a treasure trove—of courage to take on challenges, innovative, out-of-the-box thinking, and relationship-building skills. Yet these strengths often occur in the context of kids struggling to figure it all out. It's during adolescence that mental illness is most likely to emerge and the incidence of anxiety disorders reaches its peak. But the same neural circuitry underlying the anxious brain also enhances teens' ability to learn about the social world and form good relationships.

Take sixteen-year-old Marie. When she came to my lab for a study on anxiety, she found it difficult to make eye contact and answered with monosyllables my cheerful attempts to engage her. However, as she became more comfortable talking about herself, she not only told me more about the panic attacks she'd had over the past six months but she also shared a story about how her worries and "nerves" were a big part of what made her a good friend.

Marie's best friend, Sylvia, had been busy all week with hours of schoolwork and two after-school sports—a usual routine that left no time for hanging out. When she finally found time with her over a weekend milkshake, Marie saw that Sylvia kept fidgeting and looking away when she talked about the party the previous Saturday. Sylvia also put on the fake smile she used when she wanted adults to get off her case—a big red flag! As Marie watched and listened to Sylvia, her own anxiety started to rise. She just *knew* that something was wrong. So at the risk of making Sylvia mad, she pushed her to spill what was really going on. And Marie was right; it turned out that Sylvia not only had broken up with her new boyfriend but had done so because he had tried to force her to have sex with him at the party. She had just barely fought him off. Sylvia didn't know what to do or whom to tell about what had happened to her. Marie was there for her and helped her figure out the next steps to take. It was in large part thanks to Marie's anxious teenage brain that she was able to see the signs that her friend was struggling—and to give her the support she needed.

More Spice, Less Nice

When we think about youthful fragility, we may assume that girls are the most fragile of all, especially when it comes to anxiety. And it's true that although boys and girls are equally likely to show severe levels of anxiety in childhood, once puberty hits, girls are twice as likely as boys to be diagnosed with an anxiety disorder—a disparity that continues throughout women's lives. There are reams of theories and debates about where to place the blame, everything from female biology to social media. But one contributing factor is pretty much beyond debate: many girls are taught from a young age to become Little Miss Perfect.

Little Miss Perfect embodies the qualities of an ideal woman—a person who is not only smart, beautiful, and accomplished but also handy in the kitchen. She is strong but a "lady," never speaking too loudly or out of turn and flying straight instead of high. When Little Miss Perfect grows up and, through blood, sweat, and tears, breaks the glass ceiling to earn a much-coveted seat at the decision-making table, she still has to navigate conflicting expectations: she should exude confidence and strength but avoid being perceived as shrill; she should work all hours but also be a devoted family person.

The scope and intensity of these demanding and mixed messages put women squarely into the crosshairs of perfectionism, constantly on the precipice of failure—because who can live up to them? And as we saw in chapter 8, perfectionism—unlike excellencism—isn't even about striving for high achievement; it's about avoiding failure. Perfectionists believe that they are only as valuable and worthy as the perfect goals they reach, and any failure destroys that self-worth.

Unfortunately, perfectionism isn't rare among girls. An Australian study from 2006 found that 96 of 409 adolescent girls (slightly under 1 in 4) were classified as having unhealthy perfectionism. Girls from disadvantaged backgrounds are not immune to perfectionism-encouraging pressures. A 2011 study of 661 early adolescents from low-income families found that over 40 percent of the group showed high levels of self-critical perfectionism. And it runs in families. A 2020 study from the London School of Economics showed that children of perfectionistic parents—both girls and boys—were more likely to be perfectionists themselves, particularly when they learned that the love and affection of their parents—their regard—was conditional on their achievement.

How does this translate for young women today? Consider fifteen-year-old Annabelle, a "perfect" example.

Annabelle attends an academically rigorous high school and consistently sits at the top of her class. She's also one of the stars of the varsity volleyball team even though she's a sophomore, plays first-chair clarinet in the county youth orchestra, and just started dating one of the most popular boys in school.

About two months before she came to see me, things started to crumble. She was losing focus in her two hardest classes and got terrible headaches at least a few times a week. Although she studied for hours every night, she forgot half of what she read, and her grades were slipping. At home, she picked fights with her younger brother most days and spent increasingly more time alone in her room. She wasn't the only one struggling. Some girls in her grade had made a pact to self-harm together—cutting and burning themselves—after they read about it on social media. They'd asked her to join their group, telling her that cutting would make her feel better when she was really stressed and anxious, especially about homework and exams. She'd passed on the offer but hadn't ruled it out.

It's a slippery slope. Girls work hard to become Little Miss Perfect. Often they succeed and are constantly praised for their outstanding achievements—from good grades to good looks, from sweet manners to being a killer on the volleyball court. But those achievements quickly go from being outstanding to being merely the status quo. The bar for success is continually raised.

The bar for students from diverse backgrounds may be even harder to reach. Research on gifted Black girls in the United States, for example, showed that in 2012, only 9.7 percent were identified as gifted and talented, compared to 59.9 percent of White girls. Add to this underrepresentation something called *stereotype threat*—the pervasive risk of being judged according to negative societal stereotypes about their group, such as intellectual ability—and the pressure can become unbearable.

But if girls are set up to become perfectionists, that means they are also set up to become excellencists. And after all, girls on average outperform boys in most academic subjects. More high school girls than boys graduate in the top 10 percent of their classes, girls have higher overall average GPAs than boys do, and girls are more likely to take high school AP courses or honors courses than boys are. And this isn't just in the United States. A 2018 analysis of international data showed that girls outperform boys in educational achievement in 70 percent of the countries studied—regardless of the country's level of gender, political, economic, and social equality.

How can we counteract the pressures of Little Miss Perfect? Perhaps we should teach girls to take more risks? After all, we do it for our boys. Decades of research have shown that adults not only perceive that boys are less vulnerable to injuries than girls are, but they treat them that way. Observe parents on a playground with their kids who are swinging, sliding, and climbing jungle gyms. The parents are more likely to tell their sons, "You can do it!" and caution their daughters, "You need to hold tight so you don't fall." The lessons learned don't stop at the playground—or in childhood. You might have heard the statistic that men will apply for jobs that they are only 60 percent qualified for, while women won't apply until they meet almost all the qualifications. Although widely cited, this study from a single Hewlett-Packard internal report didn't explore women's reasons why. Other studies, including a report published in *Harvard Business Review* in 2014, have dug deeper. Researchers asked women and men if they had chosen not to apply for a job they were underqualified for, why not? Women were twice as likely as men to say that they didn't want to set themselves up for failure.

Applying for positions that we're not qualified for doesn't seem like the answer, but neither does holding back until we're 120 percent ready. The better solution is for us, as parents, to take a page from excellencism and help our girls—and boys—strive for excellence instead of perfection, be ready to work hard, apply for a job they *almost* feel ready for, crush the job interview, and jump off the jungle gym while they're at it. For our girls in particular, let's help them step off the Little Miss Perfect track to have more spice and less nice.

My daughter, Nandini, was a preschooler when I brought her into my lab for a research study training session. She was the guinea pig. My RAs were learning to run an experiment called "Perfect Circles." It's been around for decades. The goal is to frustrate children, then observe how they react. It seems simple—we just ask them to draw a circle, a *perfect* circle—but it's far less simple than you might think.

"Nandini, can you do something for me?" I asked. "I need you to draw a perfect green circle. Here's a crayon and paper, give it a try." Like most four-year-olds, she happily drew a circle. It was pretty good. But the task required that we tell her, "Hm . . . that's not quite right. It's a little pointy. Draw another one." Again she drew a circle and looked up expectantly, confident that she had gotten it right this time. "Hm . . . that's not quite right. It's squished here in the middle. Draw another one." This time she cocked an eyebrow at me. But she was bound and determined to get it right and drew another one. "Hm . . . that's not quite right. It's too small. Draw another one."

The experiment is timed to last exactly three and a half minutes. Time creeps along slowly and painfully when you have to tell a sweet little kid that she is doing a horrible job at one of the few things she *thought* she had mastered. Throughout those excruciating minutes, many kids dutifully keep drawing, but their frustrations leak out ("Show me how to make it better!"). Other kids get teary and distressed; we usually stop the experiment before that happens. A few kids even pretend that they're happy to keep drawing circles. Those are the people pleasers.

My daughter? Nandini kept drawing those darn circles but finally turned to me and said, "Mommy, I know I'm helping you with your research, but I think this circle is perfect enough. I think it's pretty. Can we move on soon?" There's my little excellencist in training.

It Takes Two to Tango

I didn't tell you what happened after Kavi's bike-riding lesson.

As soon as we got home, he went up to his room, clearly distressed by the experience. After a few minutes I asked him to come downstairs and join me at the kitchen table.

I took a deep breath and hit "Play" on the video recording. As we listened to it together, he saw my face go pale and my eyes become a little misty.

"What's wrong, Mommy?" he asked.

"I'm sorry, bud," I said. "I wanted you to listen to this so you could see how wrong I was. You had absolutely every reason to be scared and worried—you're new to bike riding, and you might have fallen. It's actually smart to feel anxious in situations like that. I made a big mistake when I told you it wasn't okay to be scared. I'm sorry. You did nothing wrong! And I love you just the way you are."

That last line, a mash-up of the wise words of two geniuses in their own right, Mr. Rogers and Billy Joel, worked wonders. His tense little shoulders relaxed, he looked into my eyes, and he smiled for the first time since I had pulled the Gremlin out of the garage.

The good news is that like those of all of us less-than-perfect parents, our kids' emotional immune systems can handle most of the challenges life—and we parents—throw their way. Not only that, they can thrive as a result. The good news is also that anxiety is a two-way street between parent and child—and once we discover that our kids' anxieties won't actually harm them, we might discover the same about our own.

Not long after, Kavi learned how to ride a bike. He still wobbled at times, but I was okay with that. So was he. He and I faced our anxieties together and came out stronger in the end.

10

Being Anxious in the Right Way

To learn to be anxious in the right way is to learn the ultimate.

—Søren Kierkegaard, *The Concept of Anxiety*

With this quote from the Patron Saint of Anxiety, we pick up where we began. Learning to be anxious in the right way even when it feels bad—that is the ultimate destination, and purpose, of this book.

If you've read this far, you might have a sense of what this call to action means to you—because you have, at some point in your life, run up against an inescapable fact: anxiety is hard. So hard that sometimes it doesn't merely make you feel bad, it keeps you from living the life you want.

Up until this point, I've offered few prescriptions about what you should do to be anxious in the right way. There are no to-do lists, no homework assignments, no therapeutic strategies you need to learn by heart. Nonetheless, I have made a promise to you: that if you challenge your own beliefs about anxiety—what it is, what it isn't, what it's good for, and how it impacts your life—your new mindset will change your whole experience of anxiety and, as a result, you'll create a better life and a better future.

Don't get me wrong; being a paradigm shifter won't be paint by numbers. But I do believe that changing your mindset will trigger a powerful change, helping you see the world in a fresh light, make different choices, and try new things. That will take work. If you've read this far, you've considered putting in that work.

In this last chapter, I provide three fundamental principles to help you get started on the work of befriending your anxiety. Encapsulating everything I've written thus far, they are action steps to help you stay on course when anxiety is confusing, a burden, and is getting in the way.

Notice that these are principles, not tips or strategies. That's not because strategies are bad; there are many excellent anxiety tips and tool kits out there. The problem with strategies is that they're meant to *overcome* your anxiety.

Instead, these three principles state unequivocally that overcoming your anxiety isn't the goal. The goal is to understand what your anxiety is telling you and then try to use that information to change your life for the better.

The three principles are:

1. Anxiety is information about the future; listen to it.
2. If anxiety isn't useful, let it go for the moment.
3. If anxiety is useful, do something purposeful with it.

1. Anxiety Is Information About the Future; Listen To It

Anxiety is nothing if not a dense, powerful packet of information. It combines bodily sensations—heart racing, throat tightening, face grimacing—with a stream, or sometimes a torrent, of thoughts and beliefs—worries, mental dress rehearsals, fix-it solutions—that train your attention on something important. It tells you that bad things *could* happen but haven't happened yet and that you still have the time and ability to make it right and get what you want. That's why anxiety manifests hope.

But to achieve this, anxiety *must* be uncomfortable. It needs you to sit up and pay attention to it. It's an energizing signal, boosting your focus and drive as you close the gap between where you are now and where you want to be. No other emotion so effectively keeps you trained on the future, enables you to focus on threats as well as rewards, and keeps you working toward your goals. This is why anxiety is a useful emotion: it points your whole being toward purpose.

But here's the irony: although the unpleasantness of anxiety makes you pay attention to what is important to you, its fundamental unpleasantness makes it hard to listen. Terrible feelings make us want to turn away, unless we create a habit of sitting with them before taking action to rid ourselves of them.

That's why, when it comes to listening to anxiety, curiosity is our best friend.

I don't mean you should *want* anxiety. I didn't title this book *Love Your Anxiety*—although that did cross my mind—because not all anxiety is useful. But the stance you take toward the pain of anxiety will make all the difference in terms of how uncomfortable it is, how well you can bear it, and what you can do with it.

So don't love your anxiety. Don't even like it. Just be curious about it.

This might not, at first blush, make much sense. How can you be curious about something that is hurting you? But anxiety isn't dangerous. By approaching it with curiosity, you underscore this fundamental fact. You realize that anxiety is safe to investigate. That changes everything.

Recall the Trier Social Stress Test (TSST) in chapter 1, in which socially anxious people were judged by unfriendly strangers during difficult tasks such as public speaking and tricky math tests. The participants were told ahead of time that their natural reactions to their performance anxiety—pounding heart, racing breath, a sick feeling in the pit of the stomach—were actually signs that their bodies were energized and preparing to face the tough tasks ahead. This can be hard for some people to accept because of something called anxiety sensitivity—the belief that anxiety itself is psychologically and medically harmful. But in this study, that view was corrected; the subjects were told that anxiety was healthy and would help them perform at their best, and so they were encouraged to be more curious and appreciative of the uncomfortable emotions they were about to experience.

And it worked.

Compared to the participants in the study who weren't told that anxiety is beneficial, they had healthier physical responses: their blood vessels were more relaxed, and their heart rates were slower. Since high blood pressure and a racing heart can cause wear and tear on the body over time, this shows that when the subjects stopped assuming that anxiety was harmful, it literally did less harm. Instead, their bodies responded as healthy bodies do when striving to succeed in a difficult task.

There's a second key aspect of listening to anxiety: perceiving when it grows, shrinks, or even just goes away. In other words, your anxiety level changes. It tends to peak at the beginning of a challenging time or when you hit a bump in the road, and it plummets when you overcome that setback and achieve your goal. The cessation of anxiety is just as important a piece of information as its onset. It means you can take your foot off the gas pedal. In this way anxiety is a lot like physical pain—incredibly useful when it compels you to take actions that protect your body, such as pulling your hand back from a hot pan, but just as useful when it stops, letting you know the danger has passed. Being curious about anxiety means listening to what it tells you all along the way: when it starts, when it changes, and when it goes silent.

If you want to be more open and curious about anxiety, you must rethink what you mean by the word.

To understand how, let's digress for a moment to dig into the modern language of anxiety. The word has become ubiquitous. Online text analyses show that people today are ten times as likely to write about or utter the word *anxiety* than they were forty years ago. In that sense, anxiety has become the new stress. When I was growing up in the 1980s, *stress* was the word on everyone's lips. Back then, if anyone asked, "How are you?" and my answer wasn't "Fine, thank you," there was a high probability that I would say, "Well, okay, but I'm pretty stressed." "Stressed" was shorthand for every slightly unpleasant feeling—tired, overwhelmed, angry, concerned, scared, and sad, even in the context of something joyful. How's your wedding planning going? Oh, it's great, but I'm stressed. How is your recovery from the operation going? It's pretty stressful, but I'll get through it.

Anxiety has replaced *stress* as our emotional language placeholder for every uncomfortable feeling, every sense of uncertainty; we are anxious about giving a presentation, going on a blind date, starting a new job. The word has absorbed, amoebalike, everything from dread to pleasant anticipation. Yet the mere use of the word casts our experiences in a negative light, infusing them with danger and a touch of the not quite right. In part, this is because English suffers from a lack of nuanced words for anxiety.

That isn't the case for all languages, many of which have distinct words for healthy anxiety versus that which is debilitating. In the Khmer language of Cambodia, it is not uncommon to experience fear, or *khlach*, and worry, or *kut caraeun*. In contrast, *khyal goeu*, or "wind overload," refers to an experience akin to a panic attack—a dangerous fainting spell, along with heart palpitations, blurry vision, and shortness of breath. In some Spanish-speaking cultures, *ataque de nervios*, or attack of nerves, includes uncontrollable screaming or shouting, crying, trembling, sensations of heat rising from the chest and head, dissociative or out-of-body experiences, and verbal or physical aggression. But the words for distressing anxiety and for anticipation about the future are distinct: *la preocupación* and *la ansiedad* for distress and *el afán* for eager anticipation.

I'm not claiming that Khmer- or Spanish-speaking peoples have less debilitating anxiety—although perhaps they do. Rather, it's that the word *anxiety* in the English language, which is the predominant language of the medical sciences, means everything from simple anticipation to a clinical disorder. This imprecision makes anxiety more overwhelming and difficult to pin down.

Mental health professionals from the Duke Global Health Institute, working for years in Nepal, learned firsthand how important it is to get the language for anxiety right—with unintended negative consequences when they didn't. Counselors would often translate post-traumatic stress disorder (PTSD) as *maanasik aaghaat*, or "brain shock." In Nepal, however, as well as in India and Pakistan, an important distinction is made between the brain, or *dimaag*, and the "heart-mind," or *mann*. The *dimaag* is purely physical, much like other organs, such as the lungs and heart. If the *dimaag* is damaged, it is thought to be permanent and the chance for recovery low. In contrast, if the *mann* is in distress, it is believed, the heart and mind can be helped and healed. By assigning the diagnosis "brain shock" to their patients in rural Nepal suffering from PTSD, the counselors unwittingly caused them to believe that they were untreatable, and in their distress many refused treatment. Part of the tragedy was that reimagining the language of anxiety could have prevented that painful misunderstanding.

Once you're curious about anxiety and you notice the language you use to describe it, listening to it doesn't require any complex techniques. You can rest assured that you have the ability to understand what your anxiety is telling you and that—like all emotions—it will inevitably pass. But don't miss the opportunity to lean into your feelings and thoughts—the buzzing energy like fuel coursing through your veins, the passionate want, the choking fear, the doubt often followed by a growing confidence that perhaps you do have what it takes to succeed. Feelings by their very nature are energy in need of purpose and direction. They go from anxiety to hope, from worry to wonder. You have world enough and time to be curious and observe, because you know anxiety isn't forever.

Listen to others' anxiety as well. Expressing your openness to anxiety in small ways has a big impact. If you ask your friends and family, "How was your day?" rather than a leading question such as "Did you have a good day?" the conversation changes. You enter into an investigation in which you neither assume nor wish for a particular answer. Open-ended questions don't apply pressure for cheerful, affirmative answers such as "My day was great!" Whether the answer is good or bad, worried or hopeful, you're curious about the possibilities, saying "Tell me more" or "What did that feel like?" or "I hear you." Let the feeling be; resist judging it, censoring it, or trying to engineer a solution that very moment. That will increase your ability to listen to anxiety and help your loved ones do the same.

2. If Anxiety Isn't Useful, Let It Go for the Moment

I've spent most of this book telling you not to suppress anxiety, not to fear it, and certainly not to deny or detest it. I've said that anxiety contains valuable information, and when you listen to it, you gain wisdom about yourself and what you care about. Anxiety is the emotion that can help you do what's necessary to improve your life.

But not always.

Anxiety isn't useful or straightforward *every* time. Sometimes it's slow to reveal its message. Other times it's pointless, carrying plenty of emotion but no useful information that you can discern.

This is why it's important to recognize that anxieties fall into one of two categories: anxiety that is useful and anxiety that isn't. How can you tell which is which?

You wake up thinking about that serious problem at your daughter's school, your work deadline, or the broken appliance that you *really* need to replace. You try to stop thinking about it, but your mind circles back to it incessantly. Such worries are signals telling you in no uncertain terms what's bothering you and nudging you to act in clear and specific ways.

That's useful anxiety.

Then there's anxiety that isn't useful—or useful *yet*—typically for one of two reasons: either because it leaves you with no reasonable actions you can take or because it's free floating and unattached to any clear problem. When anxiety leaves you no option, you feel out of control. You can't figure out what actions to take to alleviate your anxious feelings and resolve the situation at hand. It's like when you go to the doctor for a biopsy: there's nothing you can do until the results come back. This kind of anxiety can leave you feeling overwhelmed and helpless, stuck in a cycle of extreme worry and apprehension. It's hard to see what you can do with it.

Then there's free-floating anxiety, the sense of angst that is so vague that it's hard to identify what if anything requires your attention or what steps you should take, like when you walk around with a persistent and pernicious sense of dread, as though the world is off its axis, but for the life of you, you don't know why. Maybe, with time, the cause of this anxiety will become clear, at which point you can deal with it. Or maybe it's a false alarm—smoke but no fire. Anxiety isn't perfect. It's human, so sometimes it gets things wrong.

In either of these cases, all you can do is put your anxiety aside for safe keeping and try something different. Let anxiety go.

This doesn't mean that you should suppress it or try to erase it. Just take a break from it and go do something else. The anxiety will wait for you, and when you return to it, you might find that you've taken an action that alleviated your distress. Or maybe the anxiety wasn't really about anything after all—a misfire.

Decades of research show the best ways of letting go: cultivate experiences that slow you down and immerse you in the present. When anxiety overwhelms me, I might read a favorite poem or listen to music that transports me. I'll take a walk to enjoy the beauty of the natural world, admiring magnificent trees, noticing the play of light on a building, or focusing my attention on the exquisite veining in a leaf. I often reach out to a friend who makes me feel at peace, more like myself, because he or she knows me best in all the world.

Whatever experience works to slow you down and absorb you in the present moment, spend time there. You'll start to break the vicious cycle of anxiety that is throwing you down rabbit holes of worries and dread. You'll also gain a sense of wonder and openness that you are part of this great big universe of possibility and that you have a place in it to pursue your own special purpose.

Nurtured by these experiences, having found solace and clarity, you can turn back to anxiety later—to think about it and listen to it. You will find a way to make anxiety useful and then—the final step—do something purposeful with it.

3. If Anxiety Is Useful, Do Something Purposeful with It

We tend to approach our anxiety as though it's a failure: if you feel bad, something is wrong with you. As a result, our goal becomes to manage the anxiety so that it goes away. When you do, that's a sign you're happy and healthy.

That's exactly the opposite of what I'm suggesting.

An anxiety-free life is an impossible goal, and it's a bad idea, because you *need* anxiety to make your life better, particularly during times of challenge. As we've explored throughout this book, anxiety allows you to see what's important, focus on it while rejecting distractions, and pursue it with all your strength or fix it. It's not noise to be silenced but rather a clear, ringing signal standing out from the buzzing static of life.

Our minds spend half the time wandering. That's not an evolutionary mistake, because when a brain enters what is called the default mode during mind wandering, it's taking a rest, but it is still active; research shows that it's actually turning over thoughts about self and others, goals and options. Indeed, it's conserving energy until something grabs its attention, like a driver lazily zoning out on a country road who becomes laser focused when a snowstorm rolls up out of nowhere. Anxiety is the signal that it's time to pay attention. Its marching orders: *A storm is coming; prepare to act.*

Anxiety harnesses our attention and energy because it wants us to do something. And like any energy, which can be neither created nor destroyed, anxiety needs to be converted, channeled, given somewhere to go. Otherwise, the pressure builds up and your quality of life takes a hit.

One of the longest-running and most comprehensive longitudinal studies ever conducted, the Harvard Study of Adult Development, has enabled generations of researchers to try to determine the answer to a fundamental question: What leads to a healthy and happy life? The study began in 1938, tracking the well-being of 268 Harvard sophomores during the Great Depression—all men because women were not admitted to Harvard at the time—eventually expanding to follow more than 1,300 people from all walks of life over the course of seventy-eight years. The researchers found that along with having good relationships, one of the best predictors of health and happiness—better than social class, IQ, and genetic factors—is having a sense of purpose in life and passing it down to the next generation. This isn't so surprising; it's one of those "something your grandmother could have told you" research findings. But it's part of the reason why being anxious in the right way means channeling it toward purpose.

When my son entered seventh grade, I asked him what the word *anxiety* brought to mind. He answered, "Being alone in a room, stressed out, probably swamped with homework." When I asked my daughter, in fourth grade, the same question, she answered, "It's when you feel nervous or when you doubt you can do it. Like when you have to stand up in class to answer the teacher or do a dance onstage." Their answers did not just reflect their different personalities—and they are quite different—but more the fact that, as seventh and fourth graders, they had distinct goals and concerns. Like a compass, anxiety pointed each of them to their true north, their unique purpose—for Kavi, managing new academic demands and for Nandini navigating social impressions.

Anxiety doesn't always point us toward purpose. In obsessive compulsive disorder, anxiety drives a vicious cycle in which compulsions, like hand-washing, checking, or reassurance-seeking are all-consuming. In the moment, they dampen feelings of anxiety, but the relief is always temporary—intense anxiety returns, and the compulsions must be performed again. Compulsions don't work in the long-run because they are not purposeful, effective actions. They don't solve a problem, help us grow, or address the true circumstances of the anxiety. So the vicious cycle continues.

Yet, useful anxiety cannot be separated from purpose. That's because, as I discussed in chapter 2, anxiety is anchored in reward brain circuitry, in the dopamine-boosted motivation to persist in the face of challenge and to pursue what is pleasurable. Anxiety drives people not only to avert disaster but to achieve satisfaction, relief, hope, awe, delight, and inspiration. You're anxious only when you care, so where is your anxiety pointing you?

Anxiety pointed me toward my purpose: a career as a scientist and a writer. I could never have built a successful research lab without my anxiety-fueled abilities: being persistently curious, tirelessly pursing research puzzles, organizing like Marie Kondo, and making top-notch to-do lists, with a healthy dash of stubborn persistence and obsessive attention to detail thrown in. Anxiety has served me well as a writer, too, both in my ability to keep at a manuscript even in the twentieth revision and because I've learned that my writing is best when it's connected to the things that I care about, the things that give me a sense of purpose.

By sense of purpose I don't mean some grand vision or a burning life mission. I mean the values and priorities that make you who you are, that give your life meaning. You can explore this yourself using a technique developed by Geoffrey Cohen and David Sherman at Stanford University called *self-affirmation*. Give it a try.

Here's how: Rank the following domains according to which ones reflect values that make you who you are and make you feel good about yourself: (1) artistic skills and aesthetic appreciation, (2) a sense of humor, (3) relationships with friends and family, (4) spontaneity and living life in the moment, (5) social skills, (6) athletics, (7) musical ability and appreciation, (8) physical attractiveness, (9) creativity, (10) business and managerial skills, and (11) romantic values.

Now take your top three and write about how they reflect you and your purpose in life. Take a few minutes to explore each domain. Write until you have no words left, and then write a bit more.

Research shows that when people take time to self-affirm—to express what they hold dear and why—their moods lift, their concentration and learning improve, their relationships are more fulfilling, and even their physical health gets a boost. These benefits can persist for months or even years.

When you channel your anxiety toward pursuing and prioritizing your purpose, it becomes courage. That's when you realize it's not only okay to be anxious about something you really care about and value—it's *because* you care about it that you're anxious. That's why you keep at it, even when it's hard. Anxiety fuels your momentum, unleashes your strength. And the amazing thing about anxiety is that it will naturally diminish when you take purposeful and smart action. When you no longer need it, it steps aside.

This is why anxiety exists: it enables us to fulfill our purpose in life. Our purposes, I should say. Whether regarding family, work, hobbies, or communities of faith, people pursue an array of purposes for different reasons—some because they believe they ought to do so, others because it is the ideal for which they hope. It's important to know the difference, because whether you're motivated by oughts or ideals influences what you do next.

Take two students who have made it their purpose to get an A grade in class. One student hopes to get an A, and if he does, he will feel deeply satisfied. He is motivated to strive toward the ideals of positive achievement and growth.

The second student, in contrast, believes that getting an A is a responsibility, something he ought to achieve in order to meet his personal standards and please others. He is motivated to avoid failure and maintain a comfortable status quo.

Their motivations shape how the students pursue their

purpose. The student focused on oughts will be vigilant and deliberative, careful to avoid mistakes and follow all course requirements to the letter in order to avoid failure. The student focused on ideals, in contrast, is more likely not only to work hard but to try to exceed expectations, eager to reach beyond what is assigned in his quest to learn and achieve something new. You can see the benefits of both approaches. But which path you choose should depend on your own values.

E. Tory Higgins, a professor of psychology at Columbia University, spent decades formulating and studying how oughts and ideals influence motivation and achievement. He found that the more people pursue their purpose in ways that fit with their own personal values—that is, eagerly and expansively if they emphasize ideals; vigilantly and carefully if they emphasize oughts—the more engaged they are, the more success they have, and the better they feel about it. When there's a mismatch—when an ideals-focused person pursues a goal only because he or she "ought to," for example—that person's anxiety and distress increase. Your emotions will let you know when you're out of tune with your purpose.

Higgins and his colleagues have shown these benefits of "fit" many times over. In a study about nutritional goals, for example, people's natural emphasis on either ideals or oughts was measured prior to the experiment. Then they were urged to eat more fruits and vegetables either because of the health benefits of doing so (ideals) or because of the health costs of not doing so (oughts). When their motivation was a good fit with the reasons given to eat healthily rather than a nonfit, they ate about 20 percent more fruits and vegetables over the following week. Higgins and his colleagues showed that the benefit of fit doesn't end with healthy food; it influences what people buy, their political beliefs, and their moral judgments about right and wrong.

What about you? Are you motivated by what you feel you ought to do or by what you dream is possible? You might find that your motivations are not the same in every situation, so don't assume that you'll answer the same way every time. Your anxiety can help you figure out exactly where you stand.

I found that to be the case recently when I struggled personally with intense anxiety. I was anxious about a terribly stressful experience my husband was having at work, one that threatened his livelihood. I was trying my best to support him but was also struggling to manage my own distress.

I soon realized that it wasn't just the threat of the situation that had intensified my anxiety to what felt like an unmanageable, choking level; it was the fact that I had absolutely no control over the situation—no actions I could take, no way to really help him other than to offer support. There was nowhere for my anxiety to go because I felt as if I had no purpose.

So I shifted gears and sought out a purpose. Much of what I wanted in that instance was motivated by oughts: I desperately wished to avert disaster, make the bad stuff go away, return things to normal. But it wasn't a goal I could achieve directly. And it wasn't a good fit for my natural motivation to pursue ideals.

So I turned to the things that never fail to help me manifest my eagerly anticipated, hoped-for purpose. First, I tried to be there for my husband, offering unconditional support while seeking out my own emotional support from friends and family who could understand what we were going through. Relationships with loved ones give me an immense sense of purpose and meaning in life. Doing that succeeded in calming the fever pitch of my anxiety a bit.

Then I drew on another deeply meaningful aspect of my life: writing. I wrote down everything about the situation, told the story from every angle: the blow-by-blow sequence of events, my husband's reactions, every thought I had, and every feeling. It was not a fine piece of writing. Actually, it was terrible. But its being well written wasn't the point. Writing it out allowed me to dig deeply into what I was feeling, make sense of what was happening, and give shape to what felt like chaos. Writing enabled me to harness my anxiety to cultivate new insights and a fresh perspective. It didn't change anything or make the situation better, but for the first time in days, I felt as though I was able to handle what we were facing.

My experience is an example of how I found a path toward being anxious in the right way—even when the anxiety felt like more than I could bear. But it is also an example of privilege. I had the support of loved ones, a roof over my head, and the luxury of being able to take time to write.

Although I felt out of control, there were many aspects of my life that I could still control.

But what if there is a very real, enduring struggle, and not so many options? What if uncertainty is a constant companion and a sense of purpose is not so easy to come by? Is the notion of doing something with anxiety, of using it to pursue purpose, still useful?

I think the answer is yes, because anxiety itself isn't the burden, it's the gift that ensures that we don't give up. It is painful much of the time, but it keeps us capable of manifesting hope. People who feel only depression feel hopeless and may even give up. But people who are anxious still care about life. They still have something they believe is worth fighting for. And if they attach that care to even the smallest purpose, their anxiety will help propel them forward.

The Rescue

We live in a world of ideals and oughts. Anxiety is a partner on that journey. In this chapter, I've told you what I think you ought to do, as well as what you might ideally do, to be anxious in the right way. But it's far from easy. Change never is easy, and there's almost never one right way to do things, especially when it comes to anxiety. The multiplicity of possibilities is wonderful but also makes it hard. Luckily, there are signposts.

The biggest signpost of all is whether we are honoring anxiety—not liking it and definitely not loving it. Honoring anxiety means that we listen to it, figure out whether it's useful or not, and channel it to prioritize and pursue our purpose. When that purpose is about celebration, awe, connection, and creativity, anxiety will be a powerful engine for joy. Anxiety is prepared to do that. It evolved to do that, with our brain, body, hearts, and mind in tow.

Underneath all the things you do with your life—love your family, miss a deadline, shop at the grocery store, watch a football game with your buddies, drink a cup of tea, play the piano, survive a pandemic, work out at the gym, water-ski, scream at your kids, write poetry, go on vacation—is a deep undercurrent of anxiety, a strong, swift river with swirls and eddies that you can dip into for greater energy, wisdom, inspiration, hope, and know-how. Can you drown in such a river? Yes. But you can also ride the current forward.

No matter where you are on the spectrum of anxiety, you can listen to it and take a leap of faith that this sometimes terrifying emotion is your ally. Seeing anxiety in this way requires a perceptual shift, like the famous Rubin's vase optical illusion I mentioned before. Right before your eyes, the vase suddenly shifts from being an object in and of itself to being the negative space between two faces in profile. Which do you see, the vase, the faces, or both?

Don't rethink anxiety. Don't neutralize it. Reclaim it as you would a lost history or a forgotten gift in a box at the top of your closet. It can be a strength, and like any true strength, it contains within it vulnerabilities. Through these vulnerabilities you will find your best and truest self.

By rescuing anxiety, we rescue ourselves.

Acknowledgments

Writing this book was one of the hardest and most satisfying things I've ever done. The only thing I can compare it to was seeing my son through the treatment of his congenital heart condition, ending with open-heart surgery when he was four months old. I compare it not because writing this book was as terrible as that—far from it—but because whenever I look back on either of these two experiences, I find myself asking the same question, "How in the world did I manage that?" The answer in both cases: With (more than) a little help from my friends.

The list of remarkable people whom I am lucky to count among my friends has to begin with my agents, Richard Pine and Eliza Rothstein, and the whole Inkwell family. You're simply the best. Richard, thank you for your brilliance and humor, your kindness, and your insights which seem to be correct upwards of 92 percent of the time. Eliza, you compassionately shepherded me through many ups and downs and made the book so much better with your incisive feedback. You're the (not-so) secret weapon every single time. None of this would have been possible without you two. You took a chance on me. I'm doing my best not to screw it up.

Then there's Bill Tonelli—my shrink, professor, consigliere, and, last but not least, editor. Bill, you're one of a kind. You helped me find my voice. When I wrestled with big ideas and felt like giving up the fight, you told me, "Don't be a quitter now." I'm so damn lucky that I got to work with you.

To Karen Rinaldi and the whole team at HarperWave: Thank you for believing in the message of *Future Tense* and bringing it to the world with such excellence and wisdom. I feel so blessed to be working with you and your outstanding team.

Dr. Charles Platkin, who is both a colleague and dear friend, was my silent partner in *Future Tense*. He helped and advised me every step of the way. He is one of the most brilliant and impressive people I know, while also being a super-mensch. Thank you, Charles, for believing in me. Your generosity of spirit has made my life so much better.

Reshma Saujani and Nihal Mehta get a special shout out—I couldn't ask for two better supporters and cheerleaders. I'm so grateful for your unwavering friendship and for being the amazing people you are. You constantly inspire me. You never doubted that I could do it, so I never doubted it myself (almost never).

I thank my network of friends who are family, who shared their thoughts and stories, let me talk through all sorts of ideas, good and bad, and listened patiently to my ramblings and, at a minimum, two dozen different elevator pitches. They were there for me every step of the way: Anya Singleton and Mike Aarons, Riaz Patel and Myles Andrews,

Kim and Rob Cavallo, Raj and Laura Amin, and Nina and Rome Thomas. Love you guys. I am thankful for you every day. I'm also so very grateful to Angela Cheng Kaplan, who has transformed my life and that of my family in wonderful and numerous ways—and gave me outstanding advice about water-cooler talk. I'm working on it, Angela! You're a star.

Deep gratitude to those who generously shared their experiences and stories. Dr. Scott Parazynski, this book is elevated because I open with your story of heroism and fortitude. Drew Sensue-Weinstein, thank you for teaching me so much about anxiety and creativity. I hope you continue to share your vision with the world. David Getz, principal of East Side Middle School (MS 114), and Dr. Tony Fisher, principal of Hunter College High School, educators like you are rare and precious. I am so grateful for your commitment to the emotional health of kids and am continually inspired by your remarkable students. To all the parents and teachers I have spoken to about anxiety and emotional health across schools in New York City, including All Souls School, The Chapin School, Collegiate School, Ethical Culture Fieldston School, and The Hewitt School: I came back from every conversation with new insights and learnings. Thank you.

In writing this book, I thought a lot about kids' anxieties as they navigate this complex world. As a result, my gratitude for the people who shape my own children's lives and character has grown exponentially, notably the excellent teachers (Ms. Z—Emily Zweibel, you get special mention!) and administrators whom my children have been so lucky to have at Collegiate and Chapin. You have taught them to persist, find strength in community, to ask questions with intelligence and curiosity, and to move in the world bravely and rightly. I'm also so grateful for the parent community at our kids' schools. I cannot express how much it means that I can count on so many of you to be there for my children. The whole "It takes a village . . ." thing doesn't even touch it. Without such support and connection, it is much, much harder to leverage anxiety as the superpower it can be.

A special thank you to Tim McHenry, who is Deputy Executive Director and Chief Programmatic Officer at the Rubin Museum in Manhattan and curator of the marvelous annual Brainwave program series, where I met Dr. Parazynski among many other fascinating people. Tim, you are one of the most charming people I know, but one of those very few charming people who also possess deep kindness and wisdom. Thank you and your team for being the heart and soul of The Rubin. It is a cultural jewel, and my experiences there have shaped this book profoundly. Candy Chang and James Reeves, thank you for creating the amazing and transformative Monument for the Anxious and Hopeful, which lived and breathed at the Rubin for many wonderful months. Your art left an indelible mark on me and on this book.

I have been blessed to have an outstanding academic support team, most notably my students and colleagues at The Emotion Regulation Lab of Hunter College. None of my research mentioned in this book would have been possible without you. I'm grateful for your brilliance, persistence, and curiosity. Then there are the academics who talked with me and inspired me, even when they might not have realized it. Dr. Seth Pollak made time on his sabbatical leave to talk about the emotional lives of teens. There are few like you in the field, Seth, thank you for your transformative scientific contributions and your boundless intellectual generosity. And to my brilliant collaborators Dr. Regina Miranda and Dr. Ekatarina Likhtik, whose work has taught me so much. Regina, you pushed me to think in different ways about how thoughts and feelings intertwine, and to keep my moral center at the center of my science. Katya, your innovative research caused a revolution in how I think about anxiety. You taught me about safety—and how it's much more than the absence of threat. I am also grateful to my colleagues at The City University of New York, including the psychology department of Hunter College, The Graduate Center, and the Advanced Science Research Center. Thank you to Jennifer Raab, president of Hunter College, for providing a platform for developing and sharing the ideas in this book. I also want to thank my colleagues at NYU Langone Health, including Dr. Leigh Charvet and Dr. Keng-Yen Huang. I began my career at NYU Langone and it was there that I met my longtime collaborator Dr. Amy Krain Roy. Amy, the insights and ideas you've shared over the many years I've been lucky enough to be your colleague have shaped much of what I've written here.

I am first and foremost an emotion scientist, and I owe an immense debt of gratitude to my long-time colleagues and mentors. To my *BuDS*–Dr. Paul Hastings and Dr. Kristin Buss: I learned so much when we staged the coup and wrote the monograph. Those were the salad days. To my graduate mentor Dr. Pamela M. Cole: Pamela, you taught me that all emotions are a gift—even when they're a double-edged sword—and that culture and context matter. I am blessed to have learned about emotion, child development, and anxiety disorders from giants in the field, like Pamela, as well as Dr. Joseph Campos, Dr. Dante Cicchetti, and Dr. Tom Borkovec. Your work redefined how we understand emotional risk, well-being, and resilience. It's made the world a better place.

My research on digital technology and anxiety has been strongly shaped by my colleagues Dr. Sarah Myruski, one of the smartest people I know, Dr. Kristin Buss (again), Dr. Koraly Pérez-Edgar, and so many wonderful researchers at the Pennsylvania State University. I also thank Diane Sawyer, Claire Weinraub, and the team behind the excellent ABC Screentime Special for highlighting this work. I've also gained so much from my colleagues in the digital wellness space—especially Kim Anenberg Cavallo and the team of National Day of Unplugging, Teodora Pavkovic, Andrew Rasiej, and Micah Sifry. You understand that when technology becomes more humane, we all win.

Family saw me through. Mom and John, thank you for your love and support and for being super-grandparents. We're so lucky to have you in our lives. Aunt Beth, my godmother, you've influenced me in ways I suspect you've never guessed, including making it cool to be bookish. To the Beharrys—all of you! You've given me and our kids a tribe that is always home. Thank you to Seeta Heeralall, my right-hand woman, for being at the heart of our family for so many years. And then there's Katie and Rob Adams. The BEST. Den, you're so much more than a sister to me. Den Summit got me through some rough patches and your insights and ideas always helped me change course on *Future Tense*—and most other things—for the better. And Straw, you bring so much heart and light to our family. Thank you for being my brother.

To Noci: You were my most consistent companion throughout the years I worked on this book. Thank you for your unwavering loyalty, your calming presence, and your love of tummy rubs. During our walks I cleared my head and had many an ah-ha moment.

To Kavi and Nandini: By being in the world, you make everything better for me, more beautiful, and more hopeful. You may be a little annoyed by your appearances in the book—I hope not—but the truth is, you had to be in the book because you constantly teach me. I love you both so much.

Writing this book has been a journey of a lifetime. Thank you to my dear husband and life partner, Vivek J. Tiwary, who has given me more love and support in the writing of this book and in everything in life than I ever could have hoped for. You're my rock. By being the person you are, you have allowed me to believe that anything is possible. I love you El Capitán.

And, lastly, to our pufferfish: You encapsulate the spirit of *Future Tense*. We love you because we can see that, like all of us, you're going through some stuff. Keep swimming, my friend.

Notes

Prologue

1 "Whoever has learned": Søren Kierkegaard, *The Concept of Anxiety: A Simple Psychologically Oriented Deliberation in View of the Dogmatic Problem of Hereditary Sin*, translated by Alastair Hannay (New York: W. W. Norton, 2014), 189.

1: What Anxiety Is (and Isn't)

16 A large epidemiological study: Ronald C. Kessler and Philip S. Wang, "The Descriptive Epidemiology of Commonly Occurring Mental Disorders in the United States," *Annual Review of Public Health* 29, no. 1 (2008): 115–29, doi:10.1146/annurev.publhealth.29.020907.090847.

16 the number of Americans: "Mental Illness," National Institute of Mental Health, https://www.nimh.nih.gov/health/statistics/mental-illness.

16 Nine different anxiety disorders: *Diagnostic and Statistical Manual of Mental Disorders (DSM-5)* (Arlington, VA: American Psychiatric Association, 2017).

20 Trier Social Stress Test: Clemens Kirschbaum, Karl-Martin Pirke, and Dirk H. Hellhammer, "The 'Trier Social Stress Test'—a Tool for Investigating Psychobiological Stress Responses in a Laboratory Setting," *Neuropsychobiology* 28, nos. 1–2 (1993): 76–81, doi:10.1159/000119004.

21 In 2013, researchers at Harvard: Jeremy P. Jamieson, Matthew K. Nock, and Wendy Berry Mendes, "Changing the Conceptualization of Stress in Social Anxiety Disorder," *Clinical Psychological Science* 1, no. 4 (2013): 363–74, doi:10.1177/2167702613482119.

2: Why Anxiety Exists

27 *The Expression of the Emotions*: Charles Darwin, *The Expression of the Emotions in Man and Animals, Anniversary Edition*, 4th ed. (Oxford, UK: Oxford University Press, 2009).

27 *On the Origin of Species*: Charles Darwin, *On the Origin of Species*, vol. 5 of *The Evolution Debate, 1813–1870*, edited by David Knight (London: Routledge, 2003).

27 *The Descent of Man*: Charles Darwin, *The Descent of Man, and Selection in Relation to Sex*, vol. 22 of *The Works of Charles Darwin*, edited by Paul H. Barrett (London: Routledge, 1992).

28 "It is notorious": Darwin, *The Expression of the Emotions in Man and Animals*, 29.

29 "A man or animal driven": Ibid., 81.

29 In a classic psychology experiment: Joseph J. Campos, Alan Langer, and Alice Krowitz, "Cardiac Responses on the Visual Cliff in Prelocomotor Human Infants," *Science* 170, no. 3954 (1970): 196–97, doi:10.1126/science.170.3954.196.

30 But when the mother expresses distress: James F. Sorce et al., "Maternal Emotional Signaling: Its Effect on the Visual Cliff Behavior of 1-Year-Olds," *Developmental Psychology* 21, no. 1 (1985): 195–200, doi:10.1037/0012-1649.21.1.195.

30 Functional Emotion Theory: Karen C. Barrett and Joseph J. Campos, "Perspectives on Emotional Development II: A Functionalist Approach to Emotions," in *Handbook of Infant Development*, 2nd ed., edited by Joy D. Osofsky (New York: John Wiley & Sons, 1987), 555–78; Dacher Keltner and James J. Gross, "Functional Accounts of Emotions," *Cognition & Emotion* 13, no. 5 (1999): 467–80, doi: 10.1080/026999399379140.

30 appraisal and action readiness: Nico H. Frijda, *The Emotions* (Cambridge, UK: Cambridge University Press, 2001).

34 "But when the blast of war": Darwin, *The Expression of the Emotions in Man and Animals*, 240.

36 the release of dopamine: https://www.ncbi.nlm.nih.gov/pmc/articles/PMC3181681/

38 the defensive brain: Joseph LeDoux and Nathaniel D. Daw, "Surviving Threats: Neural Circuit and Computational Implications of a New Taxonomy of Defensive Behaviour," *Nature Reviews Neuroscience* 19, no. 5 (2018): 269–82, doi:10.1038/nrn.2018.22.

39 *threat bias*: Yair Bar-Haim et al., "Threat-Related Attentional Bias in Anxious and Nonanxious Individuals: A Meta-Analytic Study," *Psychological Bulletin* 133, no. 1 (2007): 1–24, doi:10.1037/0033 -2909.133.1.1; Colin MacLeod, Andrew Mathews, and Philip Tata, "Attentional Bias in Emotional Disorders," *Journal of Abnormal Psychology* 95, no. 1 (1986): 15–20, doi:10.1037/0021-843x.95.1.15.

41 we have documented the threat bias: Tracy A. Dennis-Tiwary et al.,

"Heterogeneity of the Anxiety-Related Attention Bias: A Review and Working Model for Future Research," *Clinical Psychological Science* 7, no. 5 (2019): 879–99, doi:10.1177/2167702619838474.

43 A few years later: James A. Coan, Hillary S. Schaefer, and Richard J. Davidson, "Lending a Hand," *Psychological Science* 17, no. 12 (2006): 1032–39, doi:10.1111/j.1467-9280.2006.01832.x.

45 "pit of despair" experiment: Harry F. Harlow and Stephen J. Suomi, "Induced Psychopathology in Monkeys," *Caltech Magazine*, 33, no. 6 (1970): 8–14, https://resolver.caltech.edu/CaltechES:33.6.monkeys.

3: Future Tense: Choose Your Own Adventure

47 "Anxiety for the future time": Thomas Hobbes, *Leviathan*, edited by Marshall Missner, Longman Library of Primary Sources in Philosophy (New York: Routledge, 2008 [1651]).

51 The majority of people think: David Dunning and Amber L. Story, "Depression, Realism, and the Overconfidence Effect: Are the Sadder Wiser When Predicting Future Actions and Events?," *Journal of Personality and Social Psychology* 61, no. 4 (1991): 521–32, doi:10.1037/0022-3514.61.4.521.

52 One example of this: Gabriele Oettingen, Doris Mayer, and Sam Portnow, "Pleasure Now, Pain Later," *Psychological Science* 27, no. 3 (2016): 345–53, doi:10.1177/0956797615620783.

53 We can see these patterns: Birgit Kleim et al., "Reduced Specificity in Episodic Future Thinking in Posttraumatic Stress Disorder," *Clinical Psychological Science* 2, no. 2 (2013): 165–73, doi:10.1177/2167702613495199.

53 Catastrophic thinking, for example: Adam D. Brown et al., "Overgeneralized Autobiographical Memory and Future Thinking in Combat Veterans with Posttraumatic Stress Disorder," *Journal of Behavior Therapy and Experimental Psychiatry* 44, no. 1 (2013): 129–34, doi:10.1016/j.jbtep.2011.11.004.

54 *pessimistic certainty*: Susan M. Andersen, "The Inevitability of Future Suffering: The Role of Depressive Predictive Certainty in Depression," *Social Cognition* 8, no. 2 (1990): 203–28, doi:10.1521/soco.1990.8.2.203.

54 it primes depression: Regina Miranda and Douglas S. Mennin, "Depression, Generalized Anxiety Disorder, and Certainty in Pessimistic Predictions About the Future," *Cognitive Therapy and Research* 31, no. 1 (2007): 71–82, doi:10.1007/s10608-006-9063-4.

54 and suicidal thinking: Joanna Sargalska, Regina Miranda, and Brett

Marroquín, "Being Certain About an Absence of the Positive: Specificity in Relation to Hopelessness and Suicidal Ideation," *International Journal of Cognitive Therapy* 4, no. 1 (2011): 104–16, doi:10.1521/ijct.2011.4.1.104.

54 helps us savor the present: Laura L. Carstensen, "The Influence of a Sense of Time on Human Development," *Science* 312, no. 5782 (2006): 1913–15, doi:10.1126/science.1127488.

55 A study did just that: Jordi Quoidbach, Alex M. Wood, and Michel Hansenne, "Back to the Future: The Effect of Daily Practice of Mental Time Travel into the Future on Happiness and Anxiety," *Journal of Positive Psychology* 4, no. 5 (2009): 349–55, doi:10.1080/17439760902992365.

56 One of the earliest studies: Ellen J. Langer, "The Illusion of Control," *Journal of Personality and Social Psychology* 32, no. 2 (1975): 311–28, doi:10.1037/0022-3514.32.2.311.

56 *internal-stable-global attributional style*: Lyn Y. Abramson, Martin E. Seligman, and John D. Teasdale, "Learned Helplessness in Humans: Critique and Reformulation," *Journal of Abnormal Psychology* 87, no. 1 (1978): 49–74, doi:10.1037/0021-843x.87.1.49.

59 To study worry: David York et al., "Effects of Worry and Somatic Anxiety Induction on Thoughts, Emotion and Physiological Activity," *Behaviour Research and Therapy* 25, no. 6 (1987): 523–26, doi:10.1016/0005-7967(87)90060-x.

62 Penn State researchers illustrated: Ayelet Meron Ruscio and T. D. Borkovec, "Experience and Appraisal of Worry Among High Worriers with and Without Generalized Anxiety Disorder," *Behaviour Research and Therapy* 42, no. 12 (2004): 1469–82, doi:10.1016/j.brat.2003.10.007.

4: The Anxiety-as-Disease Story

72 *The Divine Comedy*: Dante Alighieri, *The Divine Comedy of Dante Alighieri*, translated by Robert Hollander and Jean Hollander (New York: Anchor, 2002).

74 One of the most important books: Democritus Junior [Robert Burton], *The Anatomy of Melancholy*, 8th ed. (Philadelphia: J. W. Moore, 1857 [1621]), https://books.google.com/books?id=jTwJAAAAIAAJ.

75 "foul fiend of fear": Ibid., 163–64. Note: The second quote in the text is on p. 164.

78 *The Problem of Anxiety*: Sigmund Freud, *The Problem of Anxiety*, translated by Henry Alden Bunker (New York: Psychoanalytic Quarterly Press, 1936), https://books.google.com/books?id=uOh8CgAAQBAJ.

78 *The Age of Anxiety*: W. H. Auden, *The Age of Anxiety: A Baroque Eclogue* (New York: Random House, 1947).

79 In his report on the case: Sigmund Freud, "Analysis of a Phobia in a Five-Year-Old Boy," in *Two Case Histories ("Little Hans" and the "Rat Man")*, vol. 10 of *The Standard Edition of the Complete Psychological Works of Sigmund Freud* (London: Hogarth Press, 1909), 1–150.

80 Another famous Freudian patient: Sigmund Freud, "Notes upon a Case of Obsessional Neurosis," in Freud, *Two Case Histories ("Little Hans" and the "Rat Man")*, 151–318.

81 *Diagnostic and Statistical Manual of Mental Disorders (DSM-5)* (Arlington, VA: American Psychiatric Association, 2017).

83 the very first safe spaces: Kurt Lewin, *Resolving Social Conflicts, Selected Papers on Group Dynamics 1935–1946* (New York: Harper, 1948).

85 In it, Judith Shulevitz described: Judith Shulevitz, "In College and Hiding from Scary Ideas," *New York Times*, March 21, 2015, https://www.nytimes.com/2015/03/22/opinion/sunday/judith-shulevitz-hiding-from-scary-ideas.html.

87 A study from 2021: Guy A. Boysen et al., "Trigger Warning Efficacy: The Impact of Warnings on Affect, Attitudes, and Learning," *Scholarship of Teaching and Learning in Psychology* 7, no. 1 (2021): 39–52, doi:10.1037/stl0000150.

87 In a 2018 study: Benjamin W. Bellet, Payton J. Jones, and Richard J. McNally, "Trigger Warning: Empirical Evidence Ahead," *Journal of Behavior Therapy and Experimental Psychiatry* 61 (2018): 134–41, doi:10.1016/j.jbtep.2018.07.002.

5: Comfortably Numb

91 *The Age of Anxiety*: W. H. Auden, *The Age of Anxiety: A Baroque Eclogue* (New York: Random House, 1947).

93 Leo Sternbach changed all that: Jeannette Y. Wick, "The History of Benzodiazepines," *Consultant Pharmacist* 28, no. 9 (2013): 538–48, doi:10.4140/tcp.n.2013.538.

93 benzodiazepines were number one: Ibid.

94 2.3 billion Valium tablets: "Leo Sternbach: Valium: The Father of Mother's Little Helpers," *U.S. News & World Report*, December 27, 1999.

95 67 percent rise in prescriptions: "Overdose Death Rates," National Institute on Drug Abuse, January 29, 2021, https://www.drugabuse.gov/drug-topics/trends-statistics/overdose-death-rates.

96 the number three prescribed class: Ibid.

97 enough pills available: "Understanding the Epidemic," Centers for

Disease Control and Prevention, March 17, 2021, https://www.cdc
.gov/opioids/basics/epidemic.html.

98 Deaths involving prescription opioids alone: "Overdose Death
Rates," National Institute on Drug Abuse.

98 "reports that the pills": Barry Meier, "Origins of an Epidemic: Purdue Pharma Knew Its Opioids Were Widely Abused," *New York Times*,
May 29, 2018, https://www.nytimes.com/2018/05/29/health
/purdue-opioids-oxycontin.html.

99 18 percent of adolescents will struggle with debilitating anxiety:
"Mental Illness," National Institute of Mental Health, https://www
.nimh.nih.gov/health/statistics/mental-illness.

100 a Pew Research Center Report: Juliana Menasce Horowitz and Nikki
Graf, "Most U.S. Teens See Anxiety and Depression as a Major Problem Among Their Peers," Pew Research Center, February 20, 2019,
https://www.pewresearch.org/social-trends/2019/02/20/most
-u-s-teens-see-anxiety-and-depression-as-a-major-problem-among
-their-peers/.

101 "Bars: The Addictive Relationship": Angel Diaz, "Bars: The Addic-
tive Relationship with Xanax & Hip Hop | Complex News Pres-
ents," Complex, May 28, 2019, https://www.complex.com/music
/2019/05/bars-the-addictive-relationship-between-xanax-and-hip
-hop.

6: Blame the Machines?

107 A large-scale: Ingibjorg Eva Thorisdottir et al., "Active and Passive Social Media Use and Symptoms of Anxiety and Depressed
Mood Among Icelandic Adolescents," *Cyberpsychology, Behavior,
and Social Networking* 22, no. 8 (2019): 535–42, doi:10.1089/cyber
.2019.0079.

108 In 2010, researchers: Kevin Wise, Saleem Alhabash, and Hyojung
Park, "Emotional Responses During Social Information Seeking on
Facebook," *Cyberpsychology, Behavior, and Social Networking* 13, no. 5
(2010): 555–62, doi:10.1089/cyber.2009.0365.

110 Indeed, research shows: Ibid.

112 A few scientists have studied: Carmen Russoniello, Kevin O'Brien,
and J. M. Parks, "The Effectiveness of Casual Video Games in Improving Mood and Decreasing Stress," *Journal of Cyber Therapy and
Rehabilitation* 2, no. 1 (2009): 53–66.

112 simply scrolling through social media: Wise et al., "Emotional Responses During Social Information Seeking on Facebook."

113 Maneesh Juneja is: Maneesh Juneja, "Being Human," Maneesh Juneja,

May 23, 2017, https://maneeshjuneja.com/blog/2017/5/23/being -human.

114 The hand-holding neuroimaging study: James A. Coan, Hillary S. Schaefer, and Richard J. Davidson, "Lending a Hand," *Psychological Science* 17, no. 12 (2006): 1032–39, doi:10.1111/j.1467-9280 .2006.01832.x.

114 In 2012, researchers: Leslie J. Seltzer et al., "Instant Messages vs. Speech: Hormones and Why We Still Need to Hear Each Other," *Evolution and Human Behavior* 33, no. 1 (2012): 42–45, doi:10.1016/j .evolhumbehav.2011.05.004.

116 Some scientists argue: M. Tomasello, *A Natural History of Human Thinking* (Cambridge, MA: Harvard University Press, 2014).

116 In 2017, we explored: Sarah Myruski et al., "Digital Disruption? Maternal Mobile Device Use Is Related to Infant Social-Emotional Functioning," *Developmental Science* 21, no. 4 (2017), doi:10.1111 /desc.12610.

118 In a second study: Kimberly Marynowski, "Effectiveness of a Novel Paradigm Examining the Impact of Phubbing on Attention and Mood," April 21, 2021, CUNY Academic Works, https://academic works.cuny.edu/hc_sas_etds/714.

119 In an NPR report: Anya Kamenetz, "Teen Girls and Their Moms Get Candid About Phones and Social Media," NPR, December 17, 2018, https://www.npr.org/2018/12/17/672976298/teen-girls -and-their-moms-get-candid-about-phones-and-social-media.

120 One study: Jean M. Twenge et al., "Increases in Depressive Symptoms, Suicide-Related Outcomes, and Suicide Rates Among U.S. Adolescents After 2010 and Links to Increased New Media Screen Time," *Clinical Psychological Science* 6, no. 1 (2017): 3–17, doi:10.1177/2167702617723376.

120 Using the same data: Amy Orben and Andrew K. Przybylski, "The Association Between Adolescent Well-Being and Digital Technology Use," *Nature Human Behaviour* 3, no. 2 (2019): 173–82, doi:10.1038 /s41562-018-0506-1.

120 In one of the few: Sarah M. Coyne et al., "Does Time Spent Using Social Media Impact Mental Health?: An Eight Year Longitudinal Study," *Computers in Human Behavior* 104 (2020): 106160, doi:10.1016/j.chb.2019.106160.

121 Almost ten years after: Seltzer et al., "Instant Messages vs. Speech: Hormones and Why We Still Need to Hear Each Other."

123 a *New York Times* op-ed: Tracy A. Dennis-Tiwary, "Taking Away the Phones Won't Solve Our Teenagers' Problems," *New York Times*,

July 14, 2018, https://www.nytimes.com/2018/07/14/opinion/sunday/smartphone-addiction-teenagers-stress.html.

7: Uncertainty

127 "Uncertainty is the only": John Allen Paulos, *A Mathematician Plays the Stock Market* (New York: Basic Books, 2003).

131 And what causes: Jacob B. Hirsh and Michael Inzlicht, "The Devil You Know: Neuroticism Predicts Neural Response to Uncertainty," *Psychological Science* 19, no. 10 (2008): 962–67, doi:10.1111/j.1467-9280.2008.02183.x.

131 a 2004 meta-analysis: Sally S. Dickerson and Margaret E. Kemeny, "Acute Stressors and Cortisol Responses: A Theoretical Integration and Synthesis of Laboratory Research," *Psychological Bulletin* 130, no. 3 (2004): 355–91, doi:10.1037/0033-2909.130.3.355.

132 In one study: Erick J. Paul et al., "Neural Networks Underlying the Metacognitive Uncertainty Response," *Cortex* 71 (2015): 306–22, doi:10.1016/j.cortex.2015.07.028.

137 The science of list making: Orah R. Burack and Margie E. Lachman, "The Effects of List-Making on Recall in Young and Elderly Adults," *Journals of Gerontology: Series B: Psychological Sciences and Social Sciences* 51B, no. 4 (1996): 226–33, doi:10.1093/geronb/51b.4.p226.

139 when we need more self-control: David DeSteno, "Social Emotions and Intertemporal Choice: 'Hot' Mechanisms for Building Social and Economic Capital," *Current Directions in Psychological Science* 18, no. 5 (2009): 280–84, doi:10.1111/j.1467-8721.2009.01652.x.

139 Those feeling gratitude were willing: Leah Dickens and David DeSteno, "The Grateful Are Patient: Heightened Daily Gratitude Is Associated with Attenuated Temporal Discounting," *Emotion* 16, no. 4 (2016): 421–25, doi:10.1037/emo0000176.

140 My colleagues and I tracked: Marjolein Barendse et al., "Longitudinal Change in Adolescent Depression and Anxiety Symptoms from Before to During the COVID-19 Pandemic: A Collaborative of 12 Samples from 3 Countries," April 13, 2021, doi:10.31234/osf.io/hn7us.

140 Research from the United Kingdom: Polly Waite et al., "How Did the Mental Health of Children and Adolescents Change During Early Lockdown During the COVID-19 Pandemic in the UK?," February 4, 2021, doi:10.31234/osf.io/t8rfx.

8: Creativity

143 "Thus our human power": Rollo May, *The Meaning of Anxiety* (New York: W. W. Norton, 1977), 370.

147 Researchers first induce: Matthijs Baas et al., "Personality and Creativity: The Dual Pathway to Creativity Model and a Research Agenda," *Social and Personality Psychology Compass* 7, no. 10 (2013): 732–48, doi:10.1111/spc3.12062.

148 A 2008 study: Carsten K. De Dreu, Matthijs Baas, and Bernard A. Nijstad, "Hedonic Tone and Activation Level in the Mood-Creativity Link: Toward a Dual Pathway to Creativity Model," *Journal of Personality and Social Psychology* 94, no. 5 (2008): 739–56, doi:10.1037/0022-3514.94.5.739.

151 This relentless pursuit: Thomas Curran et al., "A Test of Social Learning and Parent Socialization Perspectives on the Development of Perfectionism," *Personality and Individual Differences* 160 (2020): 109925, doi:10.1016/j.paid.2020.109925.

151 It's called *excellencism*: Patrick Gaudreau, "On the Distinction Between Personal Standards Perfectionism and Excellencism: A Theory Elaboration and Research Agenda," *Perspectives on Psychological Science* 14, no. 2 (2018): 197–215, doi:10.1177/1745691618797940.

152 when students invest more time: Ibid.

153 We can break down: Diego Blum and Heinz Holling, "Spearman's Law of Diminishing Returns. A Meta-Analysis," *Intelligence* 65 (2017): 60–66, doi:10.1016/j.intell.2017.07.004.

153 perfectionists, counterintuitively, turn out: Patrick Gaudreau and Amanda Thompson, "Testing a 2×2 Model of Dispositional Perfectionism," *Personality and Individual Differences* 48, no. 5 (2010): 532–37, doi:10.1016/j.paid.2009.11.031.

154 perfectionists take longer: Joachim Stoeber, "Perfectionism, Efficiency, and Response Bias in Proof-Reading Performance: Extension and Replication," *Personality and Individual Differences* 50, no. 3 (2011): 426–29, doi:10.1016/j.paid.2010.10.021.

154 highly perfectionistic scientists: Benjamin Wigert, et al., "Perfectionism: The Good, the Bad, and the Creative," *Journal of Research in Personality* 46, no. 6 (2012): 775–79, doi:10.1016/j.jrp.2012.08.007.

154 In a 2012 study: Ibid.

156 A study by researchers: A. Madan et al., "Beyond Rose Colored Glasses: The Adaptive Role of Depressive and Anxious Symptoms Among Individuals with Heart Failure Who Were Evaluated for Transplantation," *Clinical Transplantation* 26, no. 3 (2012), doi: 10.1111/j.1399-0012.2012.01613.x.

158 the dizziness of freedom: Søren Kierkegaard, *The Concept of Anxiety: A Simple Psychologically Oriented Deliberation in View of the Dogmatic Problem of Hereditary Sin*, translated by Alastair Hannay (New York: W. W. Norton, 2014).

9: Kids Are Not Fragile

159 "If an anxiety": Rainer Maria Rilke, *Letters to a Young Poet*, translated by Stephen Mitchell (New York: Vintage Books, 1984), 110.

163 more than 10 million: "Mental Illness," National Institute of Mental Health, https://www.nimh.nih.gov/health/statistics/mental-illness.

163 A Pew Research Center Report: Juliana Menasce Horowitz and Nikki Graf, "Most U.S. Teens See Anxiety, Depression as Major Problems," Pew Research Center, February 20, 2019, https://www.pewresearch.org/social-trends/2019/02/20/most-u-s-teens-see-anxiety-and-depression-as-a-major-problem-among-their-peers/.

166 "you want to be the fire": Nassim Nicholas Taleb, *Antifragile: Things That Gain from Disorder* (New York: Random House, 2016), 3.

166 In 2019, 124 kids: Eli R. Lebowitz et al., "Parent-Based Treatment as Efficacious as Cognitive-Behavioral Therapy for Childhood Anxiety: A Randomized Noninferiority Study of Supportive Parenting for Anxious Childhood Emotions," *Journal of the American Academy of Child & Adolescent Psychiatry* 59, no. 3 (2020): 362–72, doi:10.1016/j.jaac.2019.02.014.

170 by the late 1990s: Howard Peter Chudacoff, *Children at Play: An American History* (New York: New York University Press, 2008).

170 snowplow parents: Claire Cain Miller and Jonah E. Bromwich, "How Parents Are Robbing Their Children of Adulthood," *New York Times*, March 16, 2019, https://www.nytimes.com/2019/03/16/style/snowplow-parenting-scandal.html.

170 2019 college admissions scandal: The Editorial Board, "Turns Out There's a Proper Way to Buy Your Kid a College Slot," *New York Times*, March 12, 2019, https://www.nytimes.com/2019/03/12/opinion/editorials/college-bribery-scandal-admissions.html.

174 massive and fundamental development: Kerstin Konrad, Christine Firk, and Peter J. Uhlhaas, 2013. "Brain Development During Adolescence: Neuroscientific Insights into This Developmental Period," *Deutsches Ärzteblatt International*, 110, no. 25 (2013): 425–31, doi:10.3238/arztebl.2013.0425.

175 A 2006 study: P. Shaw et al., 2006. "Intellectual Ability and Cortical Development in Children and Adolescents," *Nature* 440, no. 7084 (2006): 676–79, doi:10.1038/nature04513.

176 In a 2005 study: Margo Gardner and Laurence Steinberg, "Peer Influence on Risk Taking, Risk Preference, and Risky Decision Making in Adolescence and Adulthood: An Experimental Study," *Developmental Psychology* 41, no. 4 (2005): 625–35, doi:10.1037/0012-1649.41.4.625.

177 But unlike in humans: Pasko Rakic et al., "Concurrent Overpro-
duction of Synapses in Diverse Regions of the Primate Cerebral
Cortex," *Science* 232, no. 4747 (1986): 232–35, doi:10.1126/science
.3952506.

180 An Australian study from 2006: Colleen C. Hawkins, Helen M.
Watt, and Kenneth E. Sinclair, "Psychometric Properties of the Frost
Multidimensional Perfectionism Scale with Australian Adolescent
Girls," *Educational and Psychological Measurement* 66, no. 6 (2006):
1001–22, doi:10.1177/0013164405285909.

180 A 2011 study: Keith C. Herman et al., "Developmental Origins of
Perfectionism among African American Youth," *Journal of Counseling
Psychology* 58, no. 3 (2011): 321–34, doi:10.1037/a0023108.

181 A 2020 study: Curran et al., "A Test of Social Learning and Parent
Socialization Perspectives on the Development of Perfectionism."

182 The bar for students: Brittany N. Anderson and Jillian A. Martin,
"What K-12 Teachers Need to Know About Teaching Gifted Black
Girls Battling Perfectionism and Stereotype Threat," *Gifted Child
Today* 41, no. 3 (2018): 117–24, doi:10.1177/1076217518768
339.

182 Research on gifted Black girls: Civil Rights Data Collection. https://
ocrdata.ed.gov/DataAnalysisTools/DataSetBuilder?Report=7.

182 More high school girls: "2016 College-Bound Seniors Total Group
Profile Report," College Board, https://secure-media.collegeboard
.org/digitalServices/pdf/sat/total-group-2016.pdf.

182 A 2018 analysis: Gijsbert Stoet and David C. Geary, "The Gender-
Equality Paradox in Science, Technology, Engineering, and Mathematics Education," *Psychological Science* 29, no. 4 (2018): 581–93,
doi:10.1177/0956797617741719.

182 Decades of research have shown: Campbell Leaper and Rebecca S.
Bigler, "Gendered Language and Sexist Thought," *Monographs of the
Society for Research in Child Development* 69, no. 1 (2004): 128–42,
doi:10.1111/j.1540-5834.2004.06901012.x.

183 Other studies, including a report: Tara Sophia Mohr, "Why Women
Don't Apply for Jobs Unless They're 100% Qualified," *Harvard
Business Review*, August 25, 2014, https://hbr.org/2014/08/why
-women-dont-apply-for-jobs-unless-theyre-100-qualified.

183 "Perfect Circles": Elizabeth M. Planalp et al., "The Infant Version
of the Laboratory Temperament Assessment Battery (Lab-TAB):
Measurement Properties and Implications for Concepts of Temperament," *Frontiers in Psychology* 8 (2017), doi:10.3389/fpsyg.2017
.00846.

10: Being Anxious in the Right Way

187 "To learn to be anxious": Søren Kierkegaard, *The Concept of Anxiety: A Simple Psychologically Oriented Deliberation in View of the Dogmatic Problem of Hereditary Sin*, translated by Alastair Hannay (New York: W. W. Norton, 2014, [1884]).

190 Recall the Trier Social Stress Test: Jeremy P. Jamieson, Matthew K. Nock, and Wendy Berry Mendes, "Changing the Conceptualization of Stress in Social Anxiety Disorder," *Clinical Psychological Science* 1, no. 4 (2013): 363–74, doi:10.1177/2167702613482119.

193 an important distinction is made: Brandon A. Kohrt and Daniel J. Hruschka, "Nepali Concepts of Psychological Trauma: The Role of Idioms of Distress, Ethnopsychology and Ethnophysiology in Alleviating Suffering and Preventing Stigma," *Culture, Medicine, and Psychiatry* 34, no. 2 (2010): 322–52, doi:10.1007/s11013-010-9170-2.

198 because when a brain enters: Marcus E. Raichle, "The Brain's Default Mode Network," *Annual Review of Neuroscience* 38, no. 1 (2015): 433–47, doi:10.1146/annurev-neuro-071013-014030.

199 Harvard Study of Adult Development: "Harvard Second Generation Study," Harvard Medical School, https://www.adultdevelopment study.org/.

201 You can explore this: Geoffrey L. Cohen and David K. Sherman, "The Psychology of Change: Self-Affirmation and Social Psychological Intervention," *Annual Review of Psychology* 65, no. 1 (2014): 333–71, doi:10.1146/annurev-psych-010213-115137.

201 when people take time to self-affirm: Ibid.

203 E. Tory Higgins: E. Tory Higgins, "Self-Discrepancy: A Theory Relating Self and Affect," *Psychological Review* 94, no. 3 (1987): 319–40, doi:10.1037/0033-295x.94.3.319.

203 In a study about nutritional goals: Scott Spiegel, Heidi Grant-Pillow, and E. Tory Higgins, "How Regulatory Fit Enhances Motivational Strength During Goal Pursuit," *European Journal of Social Psychology* 34, no. 1 (2004): 39–54, doi:10.1002/ejsp.180.

Index

About the Author

Tracy A. Dennis-Tiwary, PhD, is a professor of psychology and neuroscience, the director of the Emotion Regulation Lab, and the coexecutive director of the Center for Health Technology at Hunter College, the City University of New York. She is also a cofounder of the digital therapeutics company Arcade Therapeutics. She has published more than one hundred scientific articles in peer-reviewed journals and delivered more than three hundred presentations at academic conferences and for corporate clients. She has been featured throughout the media, including the *New York Times*, the *Washington Post*, CBS, ABC, CNN, NPR, and Bloomberg Television. She lives with her husband and two children in New York City.